D1142218

MARS

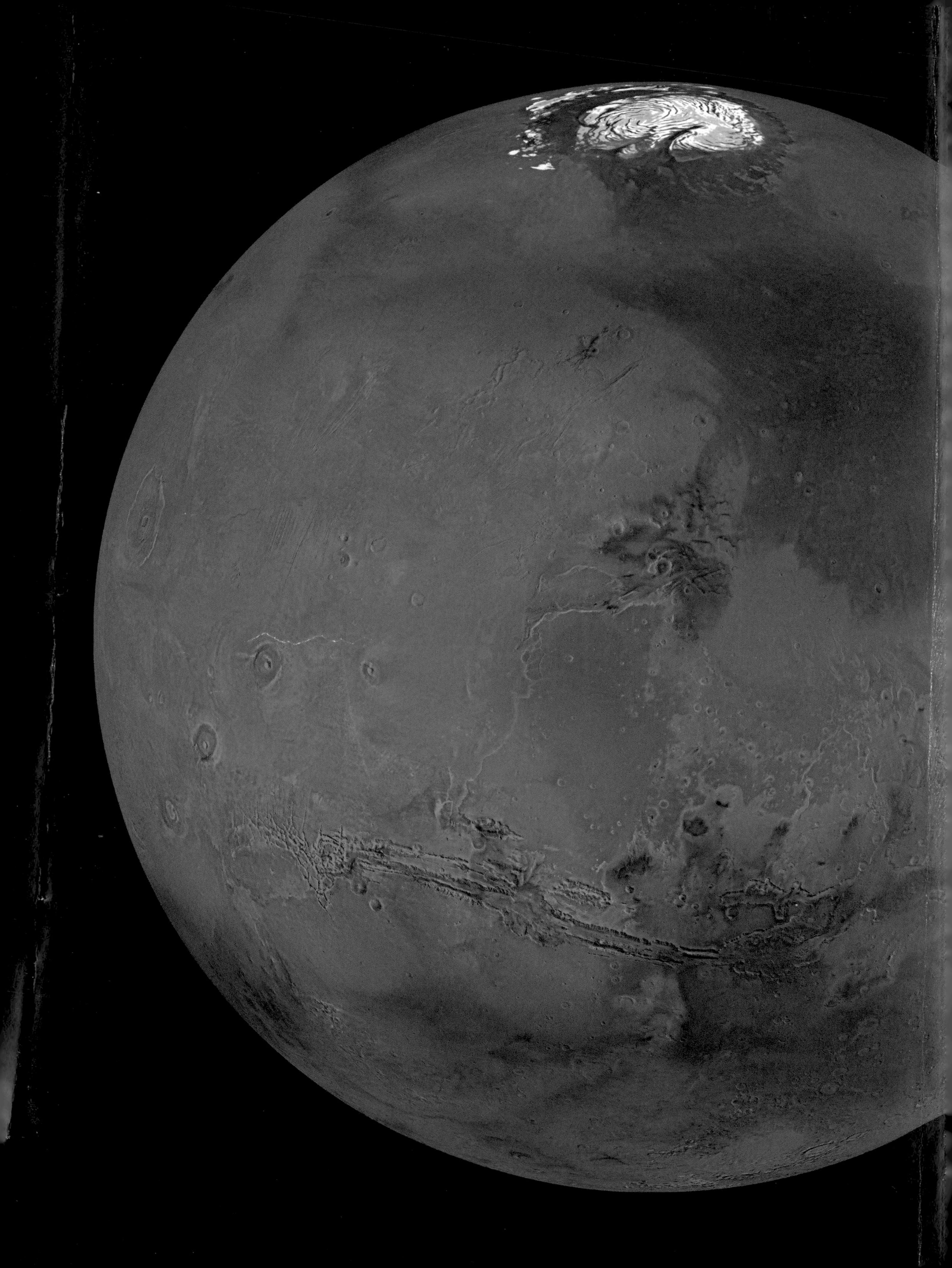

HEATHER COUPER AND NIGEL HENBEST

MARS

THE INSIDE STORY OF THE RED PLANET

HEADLINE

Also by Heather Couper and Nigel Henbest

Universe
Space Encyclopedia
The Guide to the Galaxy
To the Ends of the Universe

First published in 2001 by
HEADLINE BOOK PUBLISHING

10 9 8 7 6 5 4 3 2 1

British Library Cataloguing in Publication Data
Couper, Heather
Mars: the inside story of the Red Planet
1. Mars (Planet)
1. Title II. Henbest, Nigel
523.4'3
ISBN 0747235430

Printed and bound in Great Britain by Butler and Tanner

Designed and typeset by
Ben Cracknell Studios | Janice Mather

Headline Book Publishing
A division of Hodder Headline
338 Euston Road
London NW1 3BH

www.headline.co.uk
www.hodderheadline.com

To Rosalind Henbest
1924–2001
With love, gratitude and admiration

ACKNOWLEDGEMENTS

We're immensely grateful to all of the 50-plus Mars experts who took the time to communicate their enthusiasm and deep insights of the Red Planet to us. Too many to mention by name here, they have pride of place in the text of the book.

And we'd also like to thank the unflagging efforts of all those who helped to create such a stunning production, including our precision transcriber Nicky Ryde and the brilliant team at Headline.

CONTENTS

INTRODUCTION

In the cold mid-winter's night, Sirius and the stars of Orion scintillated like shards of ice on black velvet. Over to the east, a reddish star was shining with a steady and more homely glow. The twelve-year-old eagerly swung his newly acquired telescope towards this target: Mars, the planet of canals, vegetation and intelligent aliens...

But it was to prove a night of disappointments. The first glimpse revealed little more than a tiny ochre speck of light. Higher magnification made it swell into a tiny disc, but a disc that swirled and boiled as the beams of light from Mars were distorted by air currents in the Earth's atmosphere.

That young astronomer grew up to a career in astronomy and journalism, and became one of the authors (Nigel) of this book. And throughout that lifetime, Mars has remained frustrating.

After that first night, I kept revisiting Mars from my observing perch at the end of the wintry garden. Hopes of seeing the contentious canals were quickly dashed by the limited power of my backyard telescope. In fact, I was pleased enough just to make out some of the large dark markings on the shimmering red disc. Even here, something was wrong. My guidebooks referred to 'greenish markings, believed to be areas of vegetation'. To my eye, they were grey.

Mars was not revealing its secrets to me easily. Jupiter and Saturn were easier prey. My telescope readily revealed the swirling clouds of the Solar System's giant planet, and its four big moons. And the ringed planet hung immaculate in my field of view, like a tiny model suspended in the void. They were planets I could hold in my mind; mentally hold in my hands.

I took another tack with Mars. One of the books had a Mercator map of Mars: broad dark markings, with hints of the disputed canals. I took a ball, whitewashed it all over; drew a grid of latitude and longitude; and then carefully painted in the ochre-red deserts and the darker markings – in grey.

At last, I could hold Mars in my hand.

Meanwhile, a little girl living just outside London was also experiencing frustration with the Red Planet, for all the same reasons. In the end she (Heather) gave up with Mars, and turned her astronomical attention to stars, meteors, and – ultimately – galaxies.

But today, thanks to space technology, everything is different. This morning – almost 40 years on from our childhood disappointments – we logged on to a NASA website. Here, we can create the Red Planet on screen; turn it around; zoom in; pick out patches of ground as small as those gardens where we first observed Mars.

As children of the space age, we have both followed every twist and turn of the Martian chronicle. When we were youngsters, we believed that Mars's greenish-grey patches consisted of vegetation. Later on in school life, we experienced second-hand the shock of NASA scientists, as their

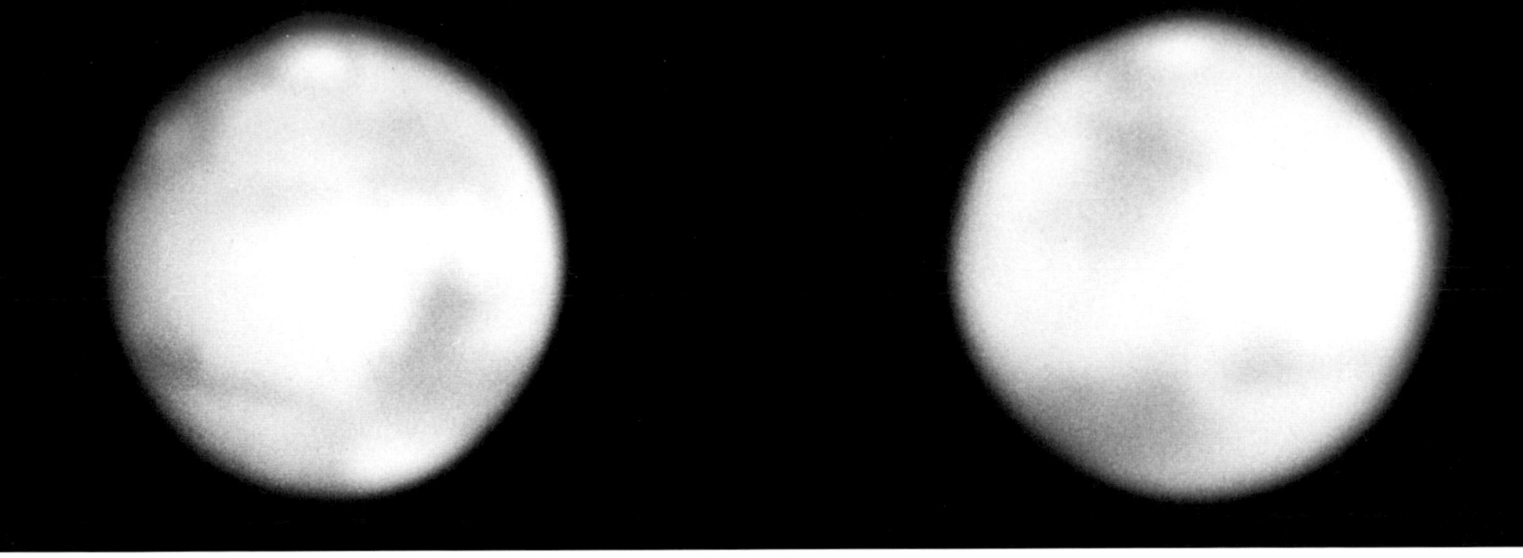

first spaceprobes revealed that Mars was pocked with craters, as dead as the Moon.

The wheel kept turning. Optimism flowered again with the discovery of dried-up river beds, hinting at ancient life on Mars. Then – following Neil Armstrong's 'small step' on the Moon – two unmanned landers touched down on the red deserts with a more ambitious mission still: to test for living cells on Mars.

At the time, we'd both been researching the far distant reaches of the Cosmos: Heather's thesis subject was 'Clusters of Galaxies' and Nigel's 'Forty-eight Extragalactic Radio Sources'. But the excitement of the search for life on Mars helped to propel us out of research into lives of communicating astronomy and science. We followed the news bulletins with the air of cognoscenti – as deeply disappointed as anyone when NASA officially declared that Mars was dead.

In search of a single simple answer, though, NASA had underestimated the Red Planet. Discoveries in the 1990s have thrown all our previous 'knowledge' of Mars into dispute. The announcement of possible 'bugs' in a meteorite from Mars was the biggest shock. At first Nigel, ever the more sceptical of us, simply refused to believe it. We had to get out there and find out what Mars is really about. So we took ourselves off around the world to talk to as many Mars experts that we could shake a microphone at – and the result is this book.

What we've discovered is a Mars that still defies all attempts to understand it. Planetary scientists wrestle with a geology that just doesn't mesh with what we see on Earth. The question of life on Mars is making scientists stand back and consider what we mean, in fact, by 'life'.

And, even as Mars seems a stranger and stranger world, we also uncovered deep links between our own world and our red neighbour. We found valuable insights into Mars coming from biologists, philosophers and science fiction writers. When we eventually tore ourselves away from the glittering trail of research and sat down at the keyboard, we realised that we weren't, after all, writing an astronomy book.

The story of Mars is a tale that penetrates deeply into our psyche. Its roots stretch back far into the past. It's a ruddy strand interwoven throughout our history, and on into our far future. Thoughts of Mars and Martians reverberate at the most basic level, raising nightmare spectres or spiritual hopes of 'others' out there in the Universe.

What we've uncovered in this quest is a Mars that's not just a rocky ball, fourth from the Sun. This ball we can now hold in our hands. But don't be taken in by its deserts and dark patches, its volcanoes and deep valleys. Scratch the surface, metaphorically, and the rocky ball becomes a crystal ball. In here we find a mirror of our minds, and the promise of our future.

BUGS FROM MARS?

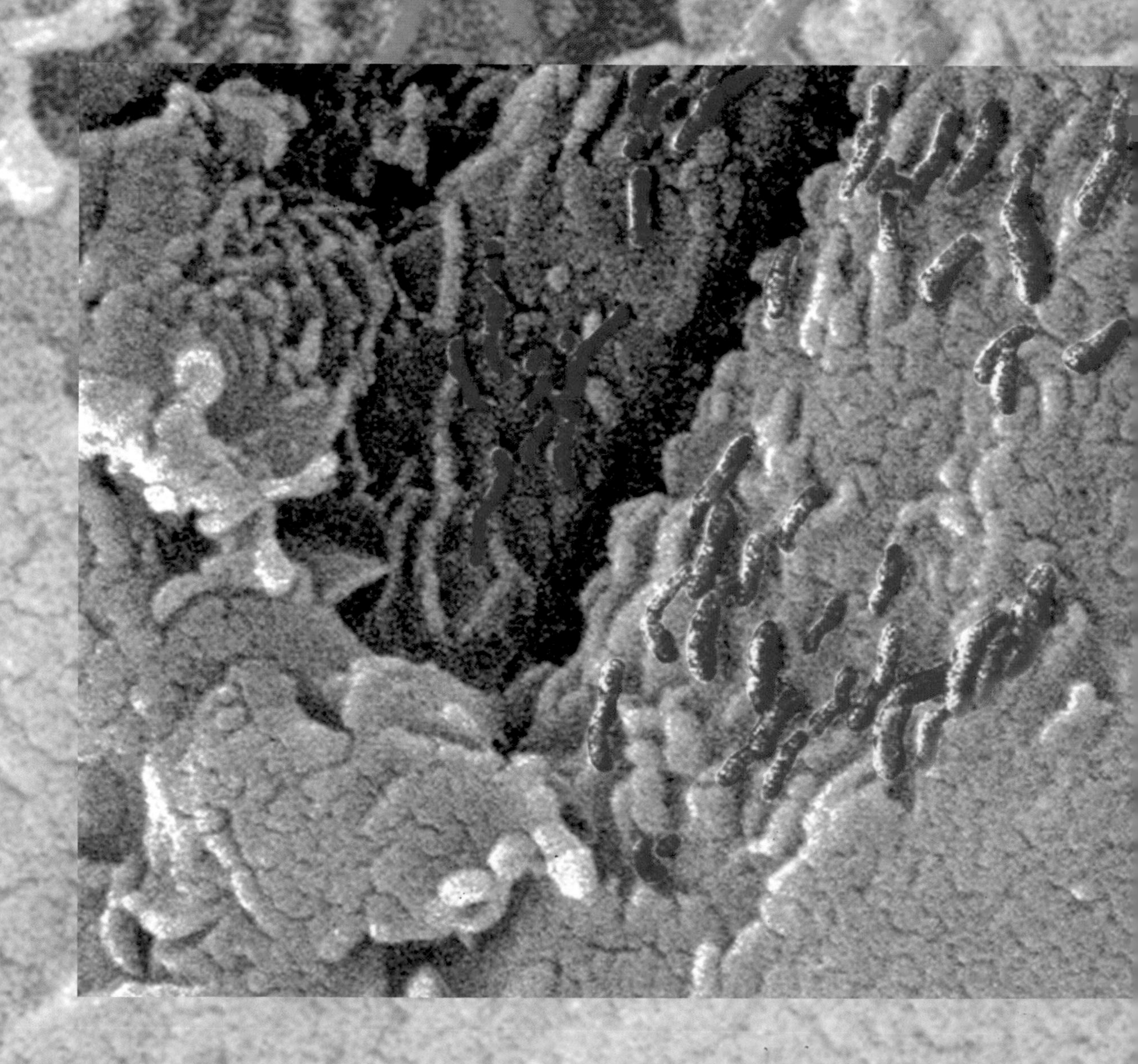

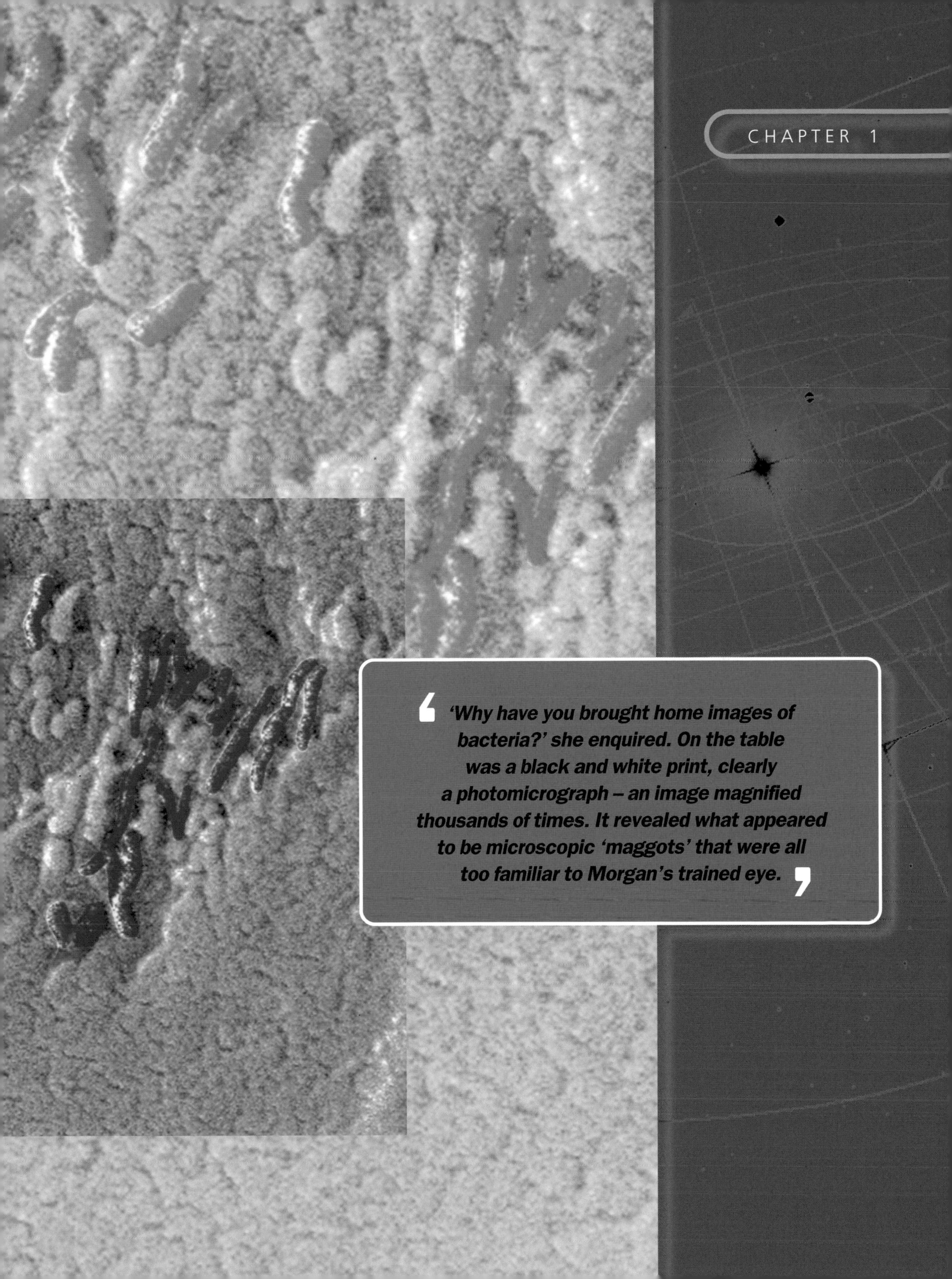

CHAPTER 1

'Why have you brought home images of bacteria?' she enquired. On the table was a black and white print, clearly a photomicrograph – an image magnified thousands of times. It revealed what appeared to be microscopic 'maggots' that were all too familiar to Morgan's trained eye.

Morgan Gibson wasn't expecting what she saw on the dining-room table that day. Her husband – an astronomer with NASA – had taken home something that brought her up rather sharpish.

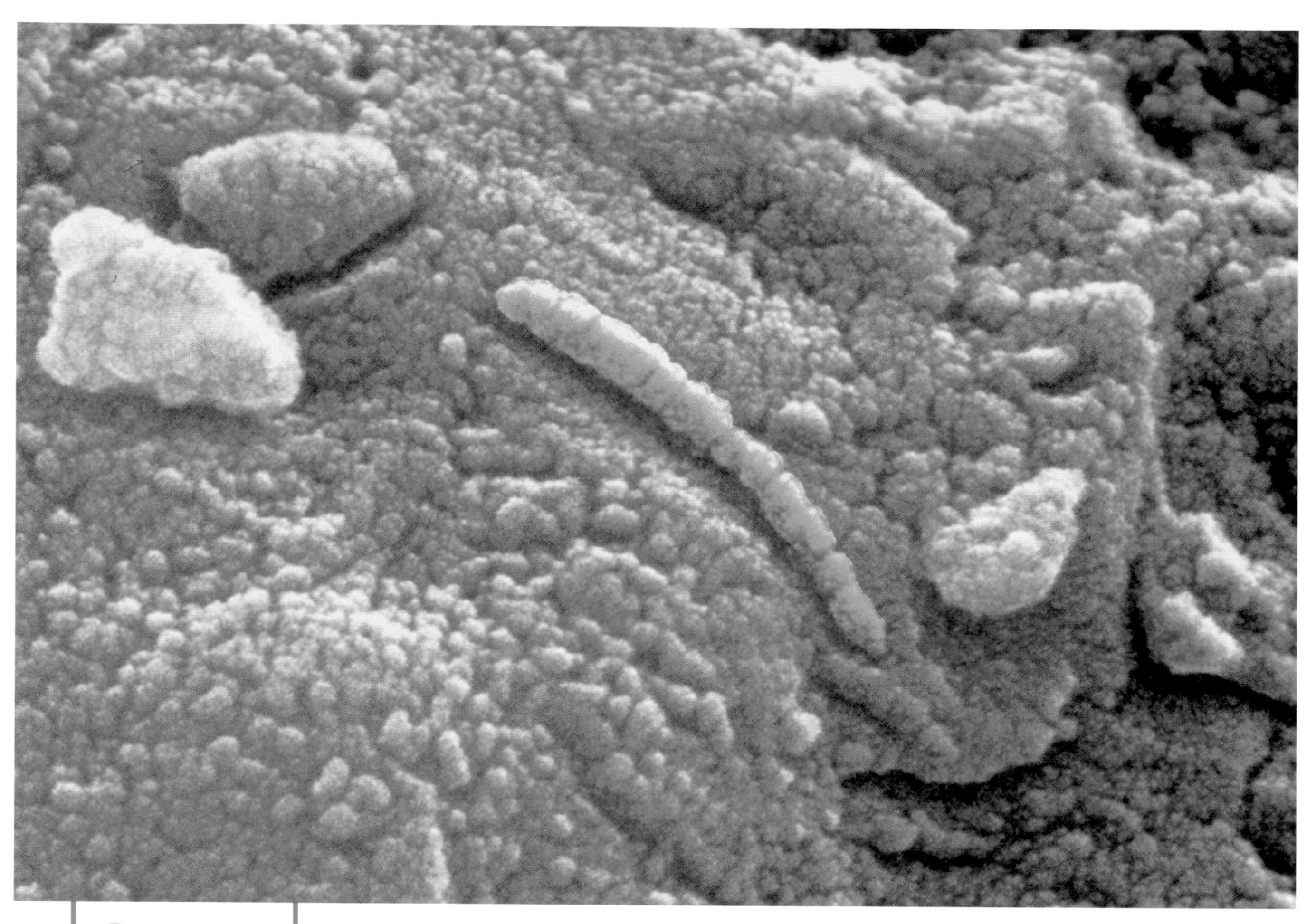

A fossilised Martian bug? Magnified thousands of times, this microscopic 'worm' looks exactly like earthly bacteria – yet it's in a rock from Mars.

Not that surprises were especially new, when you're married to Everett Gibson. Far from the caricature of a lab-coated scientist, Colonel Everett Gibson commands a vintage B-17G Second World War bomber for the 'Confederate Air Force' in his spare time. But even to Morgan, his present trophy was unusual. And, even odder, it seemed to trespass on her professional patch.

'Why have you brought home images of *bacteria*?' she enquired. On the table was a black and white print, clearly a photomicrograph – an image magnified thousands of times. It revealed what appeared to be microscopic 'maggots' that were all too familiar to Morgan's trained eye.

'That's just a rock we've been studying,' Everett responded.

'Well, it sure contains some neat bacteria,' concluded his biologist wife.

But Everett had omitted to mention that this wasn't just any old rock. In fact, it wasn't a rock from Earth at all. This rock – encapsulating the 'neat bacteria' – had come all the way from the planet Mars.

The year was 1996, and the 'bugs from Mars' were about to create waves far beyond the dining room of the Gibsons' home. From Houston, they would sweep the US and the world, as the greatest astronomical story of the decade – and possibly of all time.

> **The year was 1996, and the 'bugs from Mars' were about to create waves**

The epicentre was NASA's Johnson Space Center, a scattering of square white buildings on a huge sprawling campus to the south of the city of Houston itself. Office blocks crowd round the entrance; further off, isolated huts contain teams of scientists testing out the more hazardous elements of space flight, working daily with powerful rocket fuels such as liquid hydrogen.

For the casual visitor, the only sign of the centre's historical fame is a vast Saturn V rocket, a remnant of NASA's Moon exploration of the 1960s. It's not a mock-up but a leftover from the cancellation of the final Apollo missions. Fuelled up, this Saturn V could fly to the Moon tomorrow. But, like a monstrous beached whale, the giant rocket lies inertly on its side, beside the entrance drive from the public highway dubbed NASA Road One.

Among the anonymous buildings is Mission Control: the 'Houston' that has featured in astronaut-speak since the beginning of the American manned space programme – most famously in 1969 when Neil Armstrong announced from the lunar surface: 'Houston, Tranquillity Base here. The Eagle has landed.'

Shortly before that historic moment, Everett Gibson had applied to join the Apollo programme. At the age of twenty-seven, he was completing a research degree in Arizona and was responding to a NASA call for scientist-astronauts. Test pilots with the 'right stuff', NASA realised, were not the only people required to fly in space. With the Moon exploration coming up, they needed scientists – especially geochemists like Gibson – to interpret the alien world they would find.

'I got through several rounds of the selection,' Gibson recalls, 'but then got to the stage where my eyesight at 20/200 quickly threw me out.' So he did the next best thing, moving to Houston to work on the Moon from Mission Control. 'I've had thirty-one wonderful years here,' he says. 'Where else could you say that you were able to direct astronauts working on a foreign body, the Moon, to pick up a particular rock and collect soil samples? Then to open the boxes when they came back, to see these materials and know that you're the first person to look at this fresh material from another world.'

At the time, Gibson had no idea that his fascination with alien rocks would take him from samples of the barren Moon to rocks from a possibly living Mars. Nor did his colleague at the Johnson Space Center, Dave McKay. Unlike the extrovert Gibson, McKay is a more reticent scientist: faced with the media attention that came their way later, McKay remembers, 'There were some things that were written or were on the TV that weren't good from our point of view, because they didn't emphasise the science; they emphasised the personalities and the controversy – and we never like that. We didn't want to get into personalities.'

Dave McKay may not have wanted fame, but it was about to be thrust on him. In 1994 he got chatting with Everett Gibson about an unusual meteorite in the collection at the Johnson Space Center. Known only by its catalogue number, ALH 84001, this was no ordinary chunk of space rock; it was a piece of the planet Mars.

Gibson and his colleague Chris Romanek had found some strange chemicals in the meteorite. Under high magnification they showed up as orange 'rosettes', about five times the diameter of a human hair. The alien rosettes were made

> **'If we'd found these carbonates in a rock from Earth, we would conclude they were made by bacteria – but they were in a rock from Mars!'**

of calcium carbonate: the material of seashells. 'Chris and I came to the realisation that if we'd found these carbonates in a rock from Earth, we would conclude they were made by bacteria – but they were in a rock from Mars!'

Dave McKay had the expertise to crack down further on the mystery of orange rosettes. For years he'd been operating NASA's ultimate magnification machine, the scanning electron microscope. The researchers then recruited the final member of the NASA team that would eventually rock the world with news of 'life on Mars', a cosmic chemist called Kathie Thomas-Keprta.

'Dave, Everett and Chris invited me into the office where they sat at this huge table,' she recalls, 'and told me they thought they had found evidence for life in this meteorite – and would I take a closer look at the carbonates. At the time, I could not believe what I was hearing! And I went home that day, and told my husband: "I'm going to do this, and I'm going to prove them wrong." To begin with, I truly was the doubting Thomas of the team.'

Thomas-Keprta had already turned one branch of astronomy on its head. For fifteen

Mysterious orange 'rosettes' made NASA scientists first suspect there was something odd about Martian meteorite ALH 84001. Highly magnified here, they are no larger than this full stop.

years she had been studying tiny solid particles that drift down to Earth from space. Each microscopic dust particle is composed of thousands of minuscule grains – and scientists had always assumed the grains were made of rock. 'When you're looking at these very, very tiny grains embedded in these particles, you become quite anal-retentive about exactly what types of grains and particles you're looking at.'

'Where is everybody, why aren't they here saying, "Oh my gosh, you've found such marvellous stuff"?'

Thomas-Keprta's obsession led her to analyse the grains as never before: 'I began finding interplanetary dust particles that had thirty, forty, fifty per cent carbon – which was just amazing.' It took her ten years to convince conservative scientists that the supposed rocky fragments from space contained such a huge amount of carbon – the element of life. 'I thought, My fight's over now; I need to go on to something else. And it was just at that point that Everett and Dave approached me.'

Gibson's recollection matches hers: 'She looked at us as if we were from some distant planet – and that's not the Red Planet! – and she literally thought we were crazy.'

Thomas-Keprta's task was to home in on the tiniest grains within the Martian meteorite – little crystals of iron, oxygen and sulphur. To her experienced eye, they looked suspiciously like grains made by some bacteria on Earth. She cut the meteorite into the finest of wafer-thin slices and turned her microscope to maximum power. 'One night I was working on the microscope,' she remembers, 'and came across a particular type of iron sulphide that's *only* produced by bacteria.'

The enormity of this thought gradually sank in. 'I developed the photographs and walked outside. It was about eight, eight-thirty at night. I walked to an empty parking lot and I thought, There should be a parade out here or something. Where is everybody, why aren't they here saying, "Oh my gosh, you've found such marvellous stuff"?'

It was Thomas-Keprta's 'eureka moment' – when the original doubting Thomas swung into line with her colleagues and their audacious claim that the rock from Mars had once contained living bacteria. 'It was probably the defining moment where I had gone from being a non-believer to a believer.'

Gibson also recalls a 'eureka moment'. He and McKay were working with a microscope that yielded up 3-D views of what lay within cracks in the Martian meteorite. 'One evening David and I were on the microscope and I was looking over his shoulder – he was tweaking the knobs because he's the microscope operator. And we began looking and we came across this segmented structure. When we saw that, we both literally stopped in our tracks.'

It was a photograph of this 'segmented structure' that Everett Gibson took home and left on the dining-room table, to the surprise of his wife Morgan. 'And if a biologist looks at it and – not knowing what it is – came to that conclusion, we must have been on to something pretty significant.'

Just how significant would emerge only months later, in August 1996. NASA had arranged a press conference in Washington, to coincide with the appearance of the team's results in the respected journal *Science*. But two weeks before, with Dave McKay away camping with his family, the news began to leak...

'Before the press conference was to happen, an item appeared in *Space News* weekly,' explains Gibson, 'saying NASA had discovered

In the full glare of the world's media, Dave McKay (left) and Everett Gibson (far left) announce that ALH 84001 may have carried microfossils to Earth from Mars.

possible life on Mars. And at that time David was off on holiday and we couldn't get hold of him.' At first, NASA HQ wanted to go ahead even without McKay, but Gibson fought back: 'I told one of the associate administrators of NASA that that's not the way we release our information. I was bold enough to say, "If this is the manner you want to do it – without David and other team members present – you're talking to a *former* NASA employee".'

Fortunately, their secretary tracked down McKay at his holiday retreat, and he appeared at the press conference in Washington 'still kicking grass from his heels', as Thomas-Keprta recalls. She and Gibson were slightly ahead of him, having flown in from Houston the previous night. But none of them was prepared for what was to come...

'Our cab driver commented a big event was going to be happening at NASA that day,' says Gibson, 'and we said, "Oh, what's that?"' Coming round the corner by NASA HQ, they found the street blocked off by police. The street was packed with television relay vans sprouting antennae.

They still didn't realise the scale of the announcement until they walked into an auditorium that was crowded to overflowing. Gibson says: 'I've been through some things with Apollo and other events that were prominent, but nothing of this type ever before.' Stunned by the media presence, he leaned over to Thomas-Keprta and whispered: 'Do you realise there are thirty-four TV cameras in this room?'

She recalls: 'When Everett said that, I just couldn't look up. I swear, it was making me sick to the stomach.' To her relief, the surfeit of microphones and mini-TV cameras blew out the electrical circuits on the podium. Thomas-Keprta continues: 'I was so glad I could run and make a pit-stop – and lo and behold, I didn't realise I was two weeks pregnant and sick as a dog at that time.'

With McKay's grass shaken off his shoes, Thomas-Keprta back from the bathroom and Gibson still composed, the press conference was set to roll. And nothing was by half measures that day.

'I'm so proud of you, words can't describe it,' NASA Administrator Dan Goldin addressed the team. 'Your dedication, knowledge and painstaking research have brought us to a day that may well go down in history for American science, the American people – and indeed humanity.

'It's an unbelievable day,' NASA's chief continued in an ebullient and sparkling mood. 'Today, we're on the threshold of discovering whether life is unique to planet Earth.'

Over on the South Lawn of the White House, President Clinton was fired up too. 'Today, rock 84001 speaks to us across all those billions of years and millions of miles. It speaks of the possibility of life. If this discovery is confirmed, it will surely be one of the most stunning insights into our Universe that science has ever uncovered.'

Back at the press conference, Goldin gave the floor to the scientists. The team from the Johnson Space Center were joined by other

'Today, rock 84001 speaks to us across all those billions of years and millions of miles. It speaks of the possibility of life'

colleagues – and by a dissenter. Bill Schopf, from the University of California, Los Angeles, had discovered the oldest fossil bacteria on Earth. Goldin explained his role: 'We invited Professor William Schopf to give a point-counterpoint on the day we opened the door.'

'We held the record for the longest continuous broadcasting until Princess Diana's funeral. And nothing prepared us for what was to happen afterwards'

The Houston team had started preparing for this day even before Dave McKay had departed on holiday; Kathie Thomas-Keprta had stayed up all the previous night getting the final figures together. Their initial nervousness was betrayed only in their eyes as they calmly laid out – piece by piece – the evidence for their astounding claim.

A video animation showed blue-tinted bugs on early Mars living in a damp crack in the rock, before it was ejected into space and landed on Earth. It was an effective ploy to put the grand scenario in front of the press before the team delved into the scientific case. Thomas-Keprta showed her tiny crystals from Mars, indistinguishable from crystals made by bacteria on Earth. A colleague, Richard Zare, from Stanford University, revealed traces of organic matter in the meteorite – possibly the remains of decayed bacteria. McKay astounded the audience with his photographs of the bacteria-like shapes: 'The interpretation we favour is that these are in fact microfossils from Mars.'

If scientists had found such structures – and the associated chemistry – in an old rock from Earth, would they accept them as early fossilised cells? Gibson attempted to answer that question by consulting a list of eight criteria. Down the list, he checked one after another as positive: 'We meet a large number, if not all, of these criteria that are used to establish the presence of past life in our own terrestrial geologic column.'

Expert on early terrestrial life, and the day's nominated sceptic, Bill Schopf did not agree. Outnumbered in sheer physical terms, he gave as good as he got: 'I happen to regard the claim of life on Mars, present or past, as an extraordinary claim – and I think it is right for us to require extraordinary evidence in support of that claim.'

He didn't find the present evidence compelling. He showed a picture of the earliest bacterial fossils from Earth: they were a hundred times bigger than the claimed Martian microfossils, which could just be mineral crystals. He was sceptical that the carbonate 'rosettes' and Thomas-Keprta's crystals needed the action of living cells. And the organic matter found by Richard Zare might well have been made by chemical reactions. In conclusion, NASA's sceptic reported that he found the whole account inconclusive: 'Additional work needs to be done before we have firm confidence of life on Mars.'

But he might as well have saved his breath. The media had no time for the sceptical voice: they were after the sexy headline. The press conference itself was live on CNN television for an hour and forty minutes. Gibson believes 'we held the record for the longest continuous broadcasting until Princess Diana's funeral. And nothing prepared us for what was to happen afterwards.'

'At the end of the press conference, reporters were actually jumping over tables to get to us,' Thomas-Keprta recalls. 'We were yanked up immediately after the conference, and individual microphones and cameras were stuck in our mouths and faces,' adds Gibson. Thomas-Keprta continues: 'And then we were whisked in limousines to CNN and NBC and all over town. We didn't get back to our hotels until midnight that night. It was quite a long day and really fabulous.'

Cartoonist Kal on the *Baltimore Sun* caricatured how journalists would react to the NASA scientists' cautiously worded announcement. Judge the accuracy of his cynical view from the day's newspaper headlines (below)!

SPORTS ★ ★ ★ ★ FINAL

DAILY NEWS

NEW YORK'S HOMETOWN NEWSPAPER — Thursday, August 8, 1996

THEN THERE WERE THREE

Dole narrows veep choice

SEE PAGES 6 & 7

DAVE'S STALKER CAUGHT NEAR HIS MOM'S HOME

SEE PAGE 4

OUT OF THIS WORLD

Martian rock that brought us alien life

SEE PAGES 2 & 3

GOP deal averts abortion floor fight

COVERAGE BEGINS ON PAGES 6 & 7

America OFF-line

6 million customers get hung up as computer service goes on the fritz

NEW YORK POST

HOME DELIVERY 1-800-940-7678 CALL TODAY!

LATE CITY FINAL

MARS OR BUST!

NASA plans to send space 'armada' to Red Planet

Dr. David McKay, of the Johnson Space Center, displays Mars rock at a NASA news conference yesterday in Washington.

FULL COVERAGE ON PAGES 4 & 5

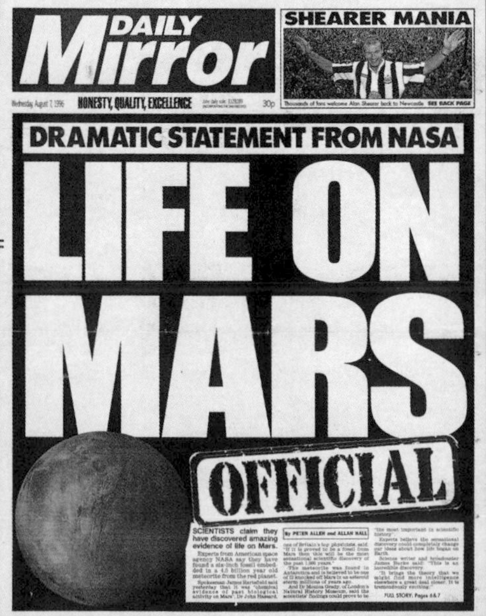

DAILY Mirror

SHEARER MANIA

HONESTY, QUALITY, EXCELLENCE — 30p

DRAMATIC STATEMENT FROM NASA

LIFE ON MARS

OFFICIAL

SCIENTISTS claim they have discovered amazing evidence of life on Mars.

But did the media get it right? 'I think the coverage was reasonably fair,' concludes McKay, 'and the media certainly did a good job of picking up on our basic points and publicising them.'

But Gibson has other views. 'The media portrayed it as if we had found life on Mars – and what we said in the conference was that we have a *chain of evidence* which could be interpreted that way.' In fact, the NASA Administrator had carefully pitched his introduction that way too: 'We have sceptical optimism.'

'It all came home to me the very next day,' recalls Gibson, 'driving to an interview in Washington when I looked out of the taxi and I saw this newsstand with three-inch-high letters: LIFE DISCOVERED ON MARS. This was not what we said, but the media had taken it to that ultimate extreme.'

And not just one headline. Around the world, the media screamed: 'Clinton hails discovery of life on Mars', 'Is this the proof of life out there?', 'Don't Panic – NASA say Martians DO exist...but 100 of them could fit on a full stop', and 'Life on Mars – Official'.

Don't Panic – NASA say Martians DO exist...but 100 of them could fit on a full stop

One person who took more than a passing interest was Monica Grady, Curator of Meteorites at London's Natural History Museum. Grady remembers her reaction to the media circus: 'I don't know whether it's my worst nightmare or my best memory ever!'

Her work on meteorite ALH 84001 has taken place in an environment very different from the 1960s complex at Houston. The Natural History Museum in London is an extravaganza of Victorian architecture, its walls a riot of variegated terracotta tiles, outside and in. The massive carved entrance doors open on to a veritable cathedral nave, ornamented in a medieval German style. Almost dwarfed by the scale of its surroundings, a massive dinosaur skeleton takes pride of place in the entrance hall.

Some of the finer details of this exuberance – carved monkeys swarming up the interior columns – engrossed us when we turned up to interview Monica Grady. She ushered us down a poky staircase to the research floor beneath the public galleries, where she inhabits a small office bulging with books and journals, with views of the perennial stream of traffic flowing along Cromwell Road.

Grady began her professional life divided between geology and chemistry. 'I loved the chemistry side; I loved the geology side, but I didn't like going out into the field in the rain.' She'd also been interested in astronomy, and an advert caught her eye for a position that required a chemist, geologist, astronomer, mathematician or physicist. 'And I thought, blimey, I'm a chemist and a geologist and I like astronomy,' she says, 'and so I got to go to Cambridge to work on lunar samples and meteorites. It kept up my geology and my chemistry – and kept me out of the rain.'

Among the thousands of meteorites she's chipped away at in her time, Grady has studied several stones from Mars. So it was natural for Dave McKay's team to contact her when they first realised that ALH 84001 was a Martian meteorite. Grady and her colleagues at Britain's Open University are ahead of the world when it comes to investigating the carbon locked up in celestial rocks.

'So we got a small chip of this meteorite, and I analysed it for its carbon, its nitrogen and its argon, and presented the results at a conference in Prague in July 1994. No suggestion of life at that time – just a new Martian meteorite that's got some very exciting features.' The next

The infamous Martian meteorite ALH 84001 – portrayed here with a 1cm cube – resembles a large potato.

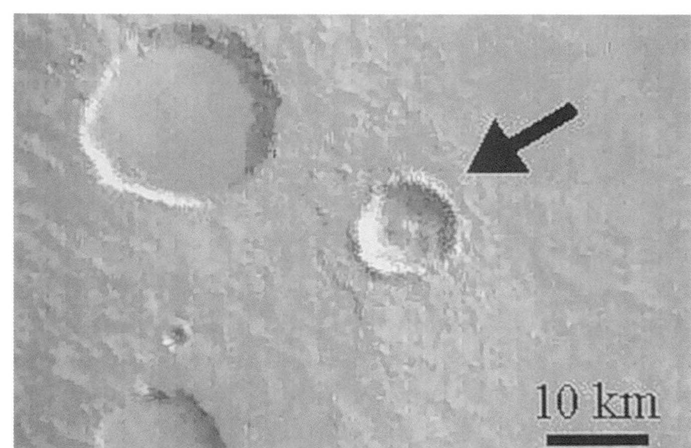

Homeland of the Martian bugs? Meteorite ALH 84001 may have been blasted into space when an infalling rock gouged out this particular crater in Mars's ancient highland region.

conference was different. It was the summer of 1996, in Berlin, and Grady remembers 'there were a lot of rumours, a lot of talking in corners about fossils'.

'So when the BBC's science correspondent rang me up one day and said, "Do you know anything about fossils?", I said, "Fossils in general, or fossils from Mars?"' Although all she'd heard at that time were rumours, Grady was game to go on the BBC and speculate. The following day saw NASA's press conference: 'And then the shit really hit the fan.' Grady was deluged with interview requests.

At the end of that week, she had a holiday booked in Cornwall. 'And it was the most disastrous bloody holiday I've ever had in my life. It rained all week, and I was just so hyper over this whole Martian meteorite thing – it was just ridiculous.'

One fact remains uncontested: ALH 84001 really is a flying rock from Mars

Back in the US, Everett Gibson also discovered that his life had changed abruptly: 'We really weren't prepared for the onslaught of things that happened afterwards. In the first year we had over a thousand enquiries for interviews, TV requests, book authors and various people, and it's quite amazing the number of things that were written, both pro and con.'

In the argument between 'pro' and 'con', Grady wasn't onside with the Houston team. 'In my humble opinion, the features that have been described as fossils I do not believe are fossilised Martian bacteria.'

The debate has raged back and forth between the experts to this day. But one fact remains uncontested: ALH 84001 really is a flying rock from Mars. The American and British teams agree on the basic history of the meteorite, using a variety of cutting-edge techniques to prise its hidden story from this apparently dumb stone.

Here's the consensus view. Rewind the history of the planets to a point some 4,500 million years ago, soon after Mars itself had formed. A stream of lava snaked down the flanks of a volcano on the tempestuous young Red Planet. The lava solidified and cracked. Mars was a warmer and wetter place then, and water seeped through cracks in the rock. The cracks filled up with the orange rosettes of the carbonates, sprinkled with Thomas-Keprta's tiny magnetic grains and McKay's bacteria-like shapes. (This much is agreed – but you have to take sides when you want to know *how* it happened: biology or chemistry?)

Fast-forward for billions of years. Just 16 million years ago – well after the dinosaurs disappeared from Earth – a small wayward planet smashed into Mars. Hitting the old lava flow a glancing blow, it blasted rocks clear of the planet altogether. The rubble was strewn across interplanetary space.

Can we tell just where on Mars this impact occurred – the home site of the famous Martian meteorite? American astronomer Nadine Barlow thinks she may have the answer. A scientist at the University of Central Florida, in Orlando, Barlow has scoured images of the whole of Mars and compiled a catalogue of 42,283 craters – scars left by the impact of asteroids and comets.

From this huge list, she's looked for craters that are oval in shape – the telltale sign of a glancing impact that could throw debris clear of Mars. Then she's eliminated all that look more than 16 million years old. 'Sixteen million years may sound like a long time to humans,' she says, 'but for geologic processes it is a very short time period.'

And in the ancient highlands of Mars, Barlow has picked out two oval craters that are sharp and clearly young. Here we have old rocks that have been blasted by a recent impact; and, in both cases, there are ancient watercourses nearby. They match neatly the history carried on board the meteorite ALH 84001. Which (if either) crater is really the home of the Martian meteorite? Barlow isn't betting: 'The only way we'll know for sure that one of these two is the actual source crater of ALH 84001, however, is to obtain actual samples of the material from these craters and analyse them. Unfortunately, that is probably still a number of years off.'

Nature liberally scattered samples of which-ever crater across the Solar System in that impact 16 million years ago. And one particular chunk eventually found itself on collision course with Mars's neighbour, the Blue Planet next in to the Sun. Some 13,000 years ago, this rock plummeted down through Earth's atmosphere as a blazing fireball, to soft-land in the snows of Antarctica. As winter followed winter, the potato-sized Martian meteorite was buried deeper and deeper in snow.

Gibson takes up the story. 'It was transported in the ice, which is always moving downslope. This ice sheet came to the Trans-Antarctic mountain range in the Allan Hills area. It tried to get over this mountain range, but it couldn't.'

Severe Antarctic winds sweep over the trapped ice, eroding the surface away at a rate of an inch or two per year. Eventually, anything that's buried in the ice will get exposed at the surface – and that includes rocks from space. NASA's Johnson Space Center has been dispatching scientists to Antarctica for many years to look out for such space debris.

'It's up to the teams that go down and walk the ice fields to pick up these objects,' Gibson continues. 'To pick up a black rock on a white ice field doesn't take a whole lot of intelligence, you know. But it does take intelligence to stay alive – and to realise that this is a rock that's different.'

At Christmas 1984, Antarctic researcher Roberta Score chanced upon a large potato-shaped rock that was rather different – it had an unusual grey-green tint. She brought it back to Houston, where it was labelled ALH 84001.

'"ALH" is for the Allan Hills site,' explains Gibson, '"84" is the field party year, 1984, and "001" means it was the first sample processed in the laboratory – not the first sample collected in the field in 1984.'

Originally, ALH 84001 was pigeonholed as an ordinary lump of space-rock, a splinter from the tiny worlds of the asteroid belt. And so it sat, mislabelled, in the NASA vaults for almost a decade, until a routine check revealed that ALH 84001 was different – especially with its strange carbonate rosettes. It was then that McKay and his team took over the investigation.

A potential alien invader comes under scrutiny: this dark stone lying on the Antarctic ice sheet could be an earthly rock, a meteorite that's fallen from space or even a rock from Mars.

First, they had to confirm their hunch that this rock should join the handful of meteorites that scientists had decided were rocks from Mars. But how can anyone distinguish a piece of Mars from an asteroid shard?

'Well, generally it's a story that comes in three instalments,' explains Grady. 'First, it's a matter of age. Most meteorites that come from the asteroid belt have ages of 4,560 million years, which is the age of the Solar System. But this odd bunch are younger, so they come from rock that was molten after the Solar System forms – so you've got to say, hey, these things come from a planet.

But how can anyone distinguish a piece of Mars from an asteroid shard?

'So we come to the second instalment: how do we tell they're not from the Moon – or even Earth,' Grady continues. The key is oxygen. Oxygen atoms come in three different kinds, and Earth and the Moon have a particular balance of the three: 'Anything on Earth shows that characteristic signature, whether it's rock, water or the atmosphere. And these particular space-rocks are different from Earth or the Moon, so that means they can only come from Mars.'

The Houston team sent a sample of the supposed asteroid fragment ALH 84001 to a lab in Chicago to test out its oxygen signature. 'When the numbers came back,' Gibson says, 'very clearly the sample had been misclassified, and it was indeed a rock from Mars.'

The third instalment in this interplanetary detective game is to check out whether the meteorite has trapped any air from its native world. This is Grady's own speciality. 'We know what Mars's atmosphere is made of,' she explains, 'because the Viking landers of 1976 measured the Martian atmosphere very accurately. The Vikings found Mars's atmosphere is 95 per cent carbon dioxide, 5 per cent nitrogen and a small amount of some unusual argon.' NASA's two Vikings were launched primarily to search for life on Mars, but their measurements are now, ironically, helping scientists to decide if there's fossil life within Martian meteorites.

Within the suspect meteorites, Grady has found chunks of black glass – blobs of rock that were melted when the meteorite was smashed off its parent world. 'And in the instant it was melted, it sucked in the atmosphere that surrounded it. If you melt those bits of glass you can get that atmosphere back – and it turns out it's got exactly the same composition as Mars's atmosphere.'

That ALH 84001 comes from Mars was perhaps the only thing that the opposing sides could agree at the time of NASA's press conference in 1996. Four years on, we set out to find how the evidence stacks up now – with the original scientists and with their colleagues who've now joined the fray.

At London's Natural History Museum, Monica Grady sticks to her guns. 'I still need to be convinced that the structure of these little features shows that they are bacteria,' she says. 'And so many of the other bits of supporting evidence that the NASA team took to show there was life in ALH 84001, they can all be explained by other reasoning. At the moment we do not have sufficient evidence for the suggestion that there are traces of life – fossils – within 84001.'

To check where the NASA team now stands, we fly from a wet summer in London to the sweltering heat of Houston. Dave McKay is out of town, but Everett Gibson and Kathie Thomas-Keprta are only too happy to join us on the veranda bar of the elegant 1960s Hilton across NASA Road One. Overlooking the busy yachts and jet skis of Clearwater Bay, it is literally an otherworldly experience to be discussing life on Mars.

Gibson has something he needs to get off his chest. 'Did you see the latest issue of *Discover* magazine?' he fumes. In a special section entitled '20 of the Greatest Blunders in Science', they have listed the Martian 'Rock of Life' along with the nuclear disaster at Chernobyl, the supposed generation of unlimited energy by 'cold fusion' and the blunders that led to the explosion of the space shuttle Challenger.

'And in the instant it was melted, it sucked in the atmosphere that surrounded it'

The article states: 'Within two years, the theory began to crack. Traces of amino acid in the rock, crucial to life, were also found in the surrounding Antarctic ice.' Gibson comments: 'Our results have got nothing to do with amino acids.'

Amino acids are the building blocks of life, simple compounds that link together to make proteins. NASA dispensed small amounts of the precious Martian meteorite to a team at the Scripps Institution in La Jolla, southern California, who have built state-of-the-art equipment that can detect even the tiniest traces of amino acids.

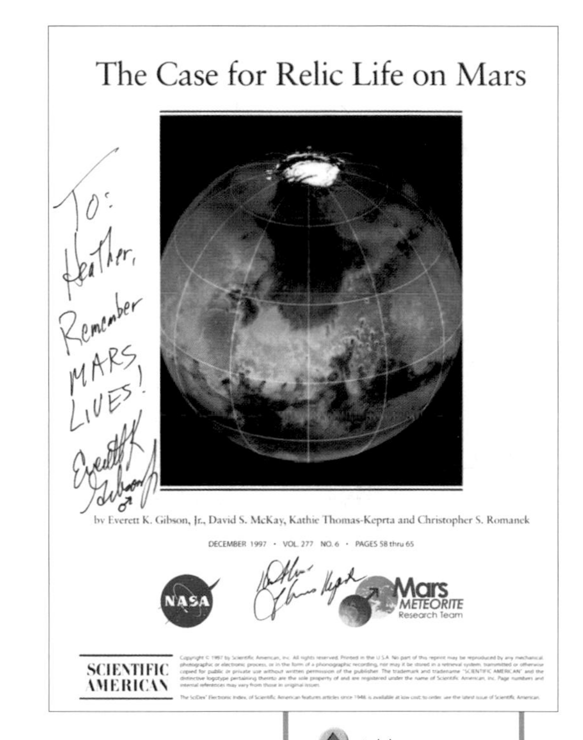

A blatant message to the authors from NASA researchers Everett Gibson and Kathie Thomas-Keprta!

Danny Glavin was a first-year research student at the time, and he recalls: 'I came in at a really good time, when we were awarded a small amount of the rock – about six pieces the size of a sugar cube. At first we were very excited to see any amino acids at all in these rocks.'

But the excitement was short-lived. All science has its checks and balances, and the Scripps team made sure they also got hold of a sample of the Antarctic ice where the meteorite had been lying. 'And unfortunately, when we analysed this ice, we found it contains the identical set of amino acids that we see in the Martian meteorite.' During the 13,000 years the meteorite spent on Earth, amino acids from the ice had seeped deep into the meteorite.

'Sure we know there's terrestrial contamination,' Thomas-Keprta continues. 'The good thing is, once we know it's there, and what it looks like, we can differentiate the terrestrial contamination from what we believe may be evidence of life that's extraterrestrial.'

That brings her back to her own pet subject, the tiny specks of minerals buried deep inside the meteorite. She's now focusing on magnetites crystals made purely of iron bonded to oxygen. 'We're talking about things that are very tiny: a billion magnetites would fit on the head of a pin.'

She convinces us that these tiny crystals cannot be contamination from the Antarctic ice. 'The magnetites are clearly embedded in the carbonate [rosettes], and we know that the carbonates formed on Mars, so we know that the magnetites formed on Mars – there's no other way.'

As we speak, Thomas-Keprta is buoyed up by new research that shows – to her complete satisfaction – that the Martian magnetites must have been made by bacteria. She hands us a thick pre-print of the paper she's about to publish in the scientific journal *Geochimica et Cosmochimica Acta*. It's bulked out by new detailed photographs of the magnetites under the microscope, compared with colourful diagrams of crystal structures. 'It's very exciting for us,' she enthuses, 'because it's the culmination of four years' worth of work. I got a bit sidelined by the birth of my son, and now we're coming out with this major paper.'

Spurred on by the Martian meteorite, Thomas-Keprta and her colleagues have gone back to make the most detailed study yet of the bacteria on Earth that produce magnetites. 'They are called magnetotactic bacteria, and we find them all over the Earth. In fact, you can go anywhere on the coast in England, say, and plunge into the murky depths and most likely you'll find certain types of magnetotactic bacteria.'

Inside these earthly bacteria, the tiny magnetites are sequestered like pearls in an oyster. To harvest a sizeable collection of the precious crystals, a team member in Iowa carefully grew magnetotactic bacteria in the lab, then prised out the cultured 'pearls'. Thomas-Keprta carefully compared these earthly magnetites with the crystals from Mars.

'It was a rather shocking – and eye-opening – comparison,' she continues. A quarter of the magnetites in ALH 84001 were totally indistinguishable from the home-grown crystals. And the team could find no way to create these pure and flawless crystals by any chemical reaction.

In addition, Thomas-Keprta points out to us a minute detail of the earthly crystals that researchers had previously overlooked. 'It was only when we found this in the Martian magnetites that we could understand the terrestrial crystals.' The correspondence seems exact. And the paper's conclusion speaks for itself: the Martian crystals were 'formed by

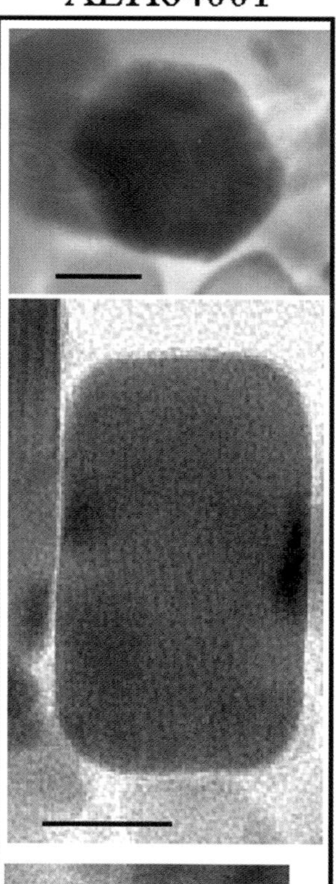

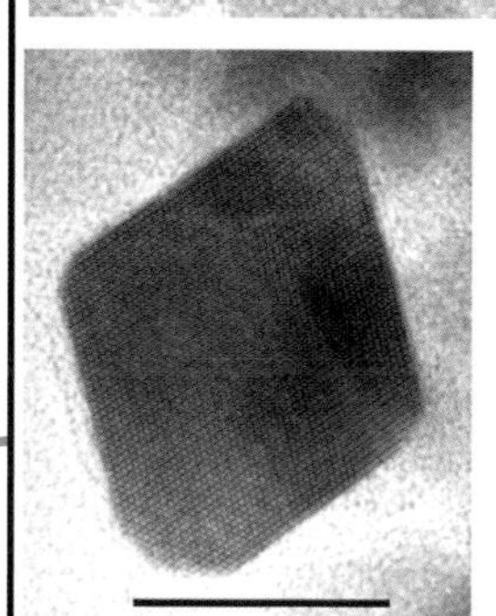

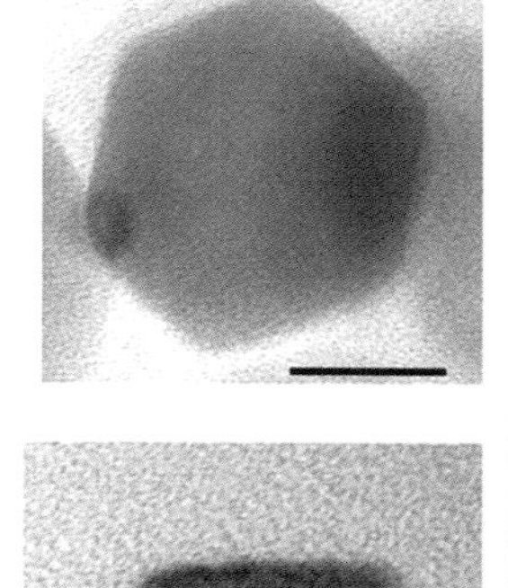

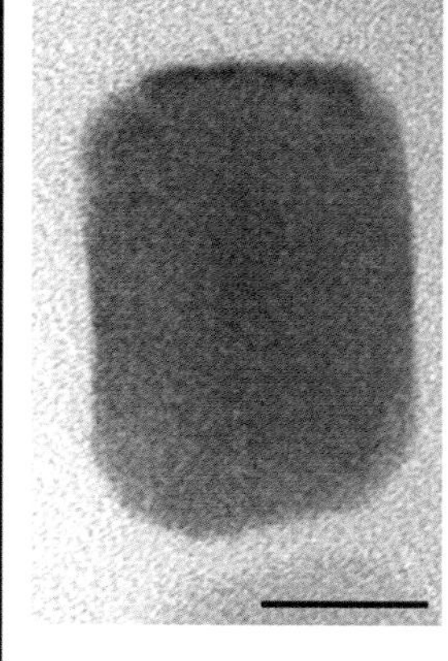

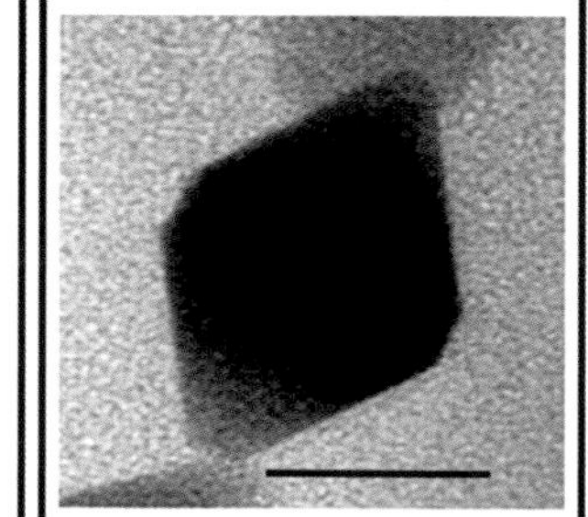

Spot the difference: magnified a mind-boggling half-million times, magnetic crystals from the Martian meteorite (left) are dead-ringers for crystals churned out by bacteria on the Earth (right).

processes similar to those that produced biogenic [life-created] magnetite crystals on Earth'.

Even so, arch sceptic Bill Schopf remains to be convinced. 'There have been four or five other detailed studies that come to absolutely the reverse conclusion. I think that the question is still unsolved as yet.'

But the evidence is beginning to sway other doubters. Bruce Murray was in charge of NASA's Viking missions to Mars in the 1970s, and throughout his career he's been highly sceptical of every supposed piece of evidence for life on the Red Planet. Thomas-Keprta's magnetites are, however, making him think again. 'I still think the odds are against them being biogenic,' Murray says. But he doesn't put the odds of Martian life at zero any more. The chance of the crystals being produced by bacteria he estimates at '20 to 25 per cent'.

'You know what they say about recognising a duck – well, this looks like a duck, it walks like a duck – and I think it quacks'

And how about those compelling pictures of the 'fossilised bacteria'? Bill Schopf is still the ultimate sceptic: 'They look nothing at all like the earliest fossilised cells on Earth,' he avers. 'And they are very much smaller than bacteria. They're the size of ribosomes – the cell's protein-manufacturing factories – and a bacterial cell on Earth has 15,000 ribosomes inside it.'

But size may not be everything. Scientists have recently been turning up tiny bugs on Earth that are also the size of ribosomes – yet are apparently alive and kicking. These 'nanobacteria' infest hot springs in Italy and the seabed off Australia – and, according to controversial research from Finland, may even live inside us as an agent that causes kidney stones.

Independent American researcher Gil Levin has certainly seen enough to be convinced. 'I've seen these nanobacteria projected on to a screen, and seen them squiggling around. So there are micro-organisms that small.'

Levin designed one of the Viking experiments that searched for living bugs on Mars in 1976, but he is now regarded as an outsider by the NASA establishment. 'While I'm an engineer, almost everything I've done has been microbial,' he expostulates. He has a photograph of an earthly micro-organism looking exactly like one of the supposed fossils in the Martian meteorite, 'and they are identical, right down to the number of segments. You know what they say about recognising a duck – well, this looks like a duck, it walks like a duck – and I think it quacks.'

And there's another, even simpler, answer to those who say the 'Martian fossils' are too small. They could be pieces that have broken off larger bacteria – perhaps fragments of hair-like filaments that they use to swim around.

Updating us on the Houston veranda bar, Everett Gibson hints at new evidence that may put all the doubts to rest. Within a different meteorite from Mars, 'we've found micro-fossil-like features that are much larger than those in 84001. It's very ironic that if our team had looked at this one first, I think it would have been easier in selling our case.'

This meteorite is called Nakhla. Along with ALH 84001, it's one of some seventeen Martian meteorites now known. While 84001 was found long after it fell to Earth, Nakhla was seen blazing its way through the sky en route from Mars to the surface of our planet.

The year was 1911. The country was Egypt. Villagers in the remote village of Nakhla were shocked as a blazing fireball swept through the sky, dropping a shower of stones behind it.

There's an urban myth among astronomers that a dog was killed by the Nakhla meteorite – the only death attributable to a stone from space. We checked out this real-life version of *Mars Attacks* with Monica Grady at the Natural History Museum.

'One contemporary report,' she confirms, 'does allege that a dog was killed by this thing – completely burned to ashes in an instant.' A stone from space may be hot when it falls, but it's not going to cause sudden incineration on this scale. 'And when you look into the archives,' Grady continues, 'you find that this particular village wasn't on the track of the meteorite fall – and they reported the death of this dog a couple of days before the meteorite actually fell. So we reckon it's someone trying to claim something exciting which didn't really happen.'

To reach out and touch Mars! We gently cradle the rough surface of Nakhla, holding part of that remote planet we've both observed so many times through a telescope

The fragments of the Nakhla meteorite were collected within days, and some of them were lodged at London's Natural History Museum. It suddenly strikes us that we're sitting only a few yards of a piece of Mars. To see a piece of the planet at first hand suddenly feels an essential prerequisite to writing this book.

We may be tentative in asking Monica Grady for a sight of a Martian meteorite, but she's only too happy to oblige – after collecting the keys that will allow her access to the locked vault. She emerges with two small chips of rock. They're in small clear plastic cases, like transparent jewellery boxes. For us, as lifelong astronomers, these rocky splinters are more moving than the Crown Jewels.

Without prior knowledge, you wouldn't look twice. One box apparently contains an ochre sugar cube, with subtle green highlights. The other holds a smaller pyramid, shining black and grey. The first is part of the Nakhla meteorite; the second a tiny chip from ALH 84001. 'May we touch them?' 'Sure,' replies the curator.

To reach out and touch Mars! We gently cradle the rough surface of Nakhla, holding part of that remote planet we've both observed so many times through a telescope. The splinter of 84001 has a sharp end. Involuntarily, I withdraw my finger from its tip, lest it draw blood: subconsciously my mind is recalling the bacteria it may contain!

Just as NASA provided this splinter of ALH 84001 to the Natural History Museum, Grady has sent samples of Nakhla to the Houston team. It has one huge advantage over ALH 84001; instead of sitting around for thousands of years on Earth, picking up contamination, Nakhla was immediately whisked into the sterility of a museum environment.

'We opened it in a clean lab,' Everett Gibson recalls, 'and inside it found samples of clays which are probably in the 600- to 700-million-year-old time frame. And within these carbonates and clays are structures and features that are even larger and better preserved than those we saw in 84001.'

And that's not all. In the 1860s a Martian meteorite fell near the Indian village of Shergotty. And the Houston team has discovered more large 'structures' resembling fossilised bacteria inside the Shergotty meteorite. 'So we have three samples now,' says Gibson, 'one that's old, one that's in the middle life of the planet and one – Shergotty – that's only 165 million years old.'

And these three specimens are far from exhausting the treasure trove of meteorites from

Mars. One fell in France, in the small village of Chassigny. 'It was towards the end of the Napoleonic Wars,' Grady muses, 'and I can imagine it must have been like the rattle of gunfire as it fell on people's roofs, and imagine them thinking, Oh God, the Napoleonic army's come back.'

came from space; they were fragments of the planet Mars. 'They realised that this man had brought in – off the street – 700 grammes of Martian material,' Everett Gibson chuckles, 'and the going price for Martian meteorites is a considerable sum of money. A fair market value for these two is probably a million dollars.

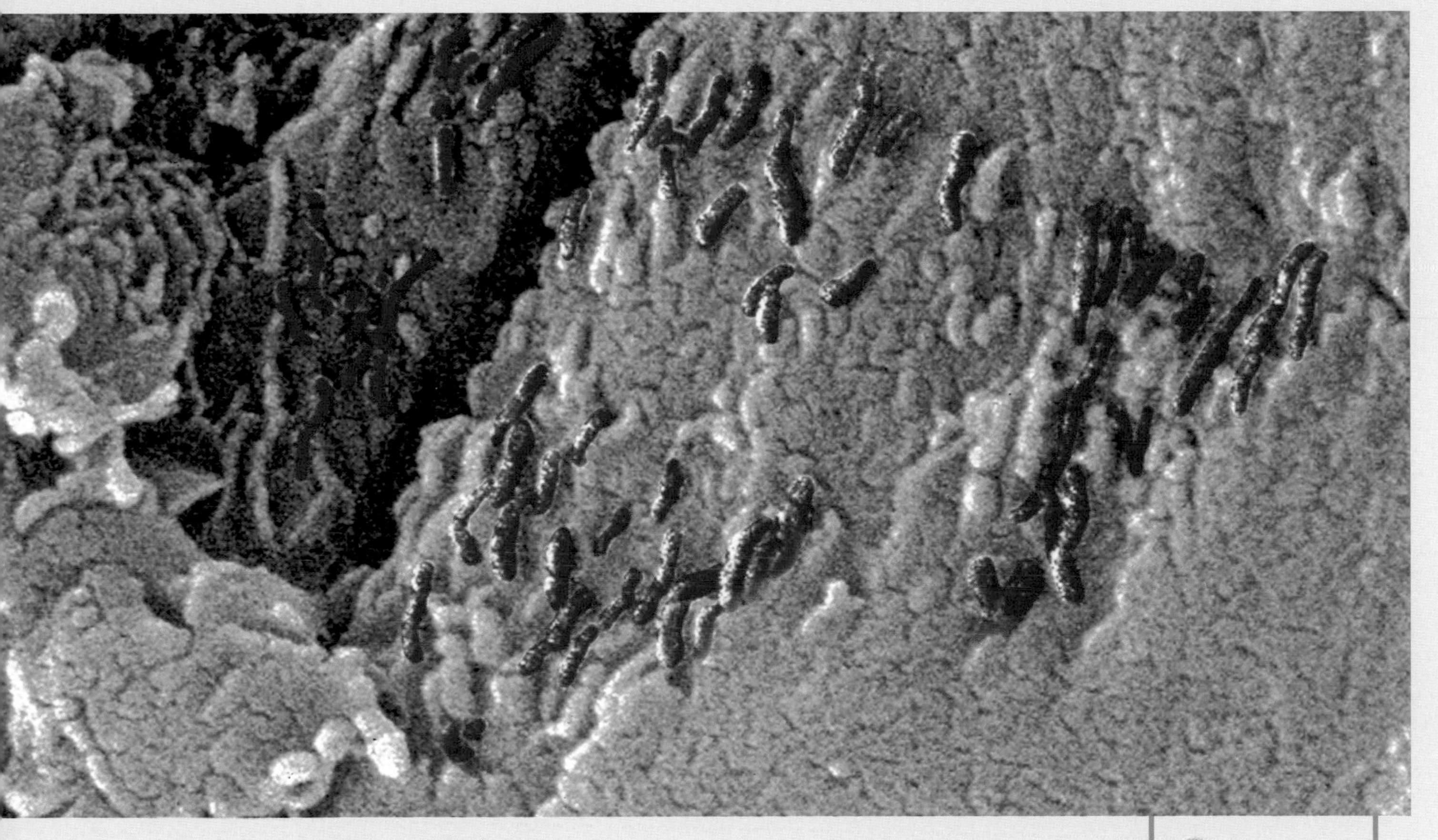

Alien microbes or our earliest ancestors? Relatives of these supposed bug-fossils (coloured blue) in the Martian meteorite may have been the seeds of life on other planets – including Earth

A very different thought was in rock-hound Bob Verish's mind as he walked California's Mojave Desert in around 1980. His eye was caught by two odd-coloured rocks, the size of small bricks. They joined his rock collection, stacked in plastic milk crates in his back yard. Years later, Verish's interest turned to meteorites. It stirred vague memories, and – sorting through the fifty or sixty milk crates he'd then filled – he eventually unearthed the twin rocks from the Mojave.

Researchers at the University of California, Los Angeles, confirmed that Verish's rocks not only

'Bob's wife said, "Sell them",' Gibson continues, 'and the scientific community said, "Preserve them". And as fate would have it, he took one of the stones and made it available to the scientific community to study.'

Surely this plethora of meteorites will lead to more scientists investigating them for signs of life? It's not that simple, Kathie Thomas-Keprta responds. 'I believe we are the only group that continues to look at the Martian meteorites themselves.'

Their concerted assault on ALH 84001, in particular, has made it the most-studied rock in the entire history of geology. Many of their critics are trying to catch up, by investigating earthly rocks in the same intimate detail – to check if they contain worm-like 'fossils' or magnetites that have clearly been made without the action of living cells.

Whatever the future brings with the Martian rocks, it's going to be a long slow campaign. For every piece of evidence that the Houston team unearth that is 'pro' life on Mars, the opposition voices will find an explanation that doesn't involve living bacteria. Kathie Thomas-Keprta is not dismayed. 'This work can be very frustrating, but you can't be constantly on the defensive or you're never moving forward. And the flavour of this group is that we're always moving forward.'

But there's one thing that everyone agrees on. The announcement in 1996 gave a kick-start to the search for life on Mars – and to the search for life in the Universe at large.

At NASA's headquarters in Washington DC, Scott Hubbard is the former head of Mars exploration. He sees the meteorite ALH 84001 as having a far wider impact than opening a fresh perspective on the Red Planet: 'I think that the finding in 1996 was a pivotal moment in focusing NASA's space science endeavour on what has come to be called the Origins Initiative.'

As we researched this book, we felt a new thrill running through the community of scientists who study the Red Planet

It led NASA to create an Astrobiology Institute across the continent in California, and to appoint as its director not an astronomer but a human biologist. Baruch Blumberg won the Nobel Prize for his work on the liver disease hepatitis B, and now he spearheads NASA's search for life beyond. His assessment of the importance of ALH 84001 is forthright: 'The debate that rock started, and the subsequent excitement, was the major impetus to the whole astrobiology programme.'

And where will it lead? In the long run, the meteorites from Mars have provided a spur to mankind's long-held ambition to reach out for the Red Planet. Bruce Jakosky, at the University of Colorado at Boulder – nestling under the Rocky Mountains – has long been a fan of Mars. 'My sense is that the data at best are ambiguous and equivocal, and we're not going to find an answer from these couple of rocks. If you picked up a random rock on the Earth you wouldn't find evidence of life necessarily; so we've picked up a random rock from Mars and we shouldn't take that as the final answer on this question. The only way to answer the question about whether there has been life on Mars is to go to Mars.'

As we researched this book, we felt a new thrill running through the community of scientists who study the Red Planet. It's partly, of course, the tantalising possibility of extinct life in ALH 84001 – the 'Rosetta Stone of the history of planet Mars', in Everett Gibson's words. And it's partly the growing surge of momentum towards sending robot craft to Mars, and ultimately a human mission to explore Mars.

But there's a growing realisation of a much deeper bond between us humans and the planet Mars.

'One of the things that we must remember,' says Martian-meteorite expert Everett Gibson, 'is that the Earth has had Martian material arriving at it over the entire history of our planet. The planets regularly exchange material among themselves.'

As infalling asteroids pound the surface of Mars, they eject sprays of rock that may fall on Earth. And rocks blasted off Earth may wing their way through interplanetary space and end up on

If life began on Mars, then the 'bugs' we are...seeing in the Martian meteorite ALH 84001 are...in fact our own earliest ancestors

At stately Stanford University, a quiet oasis in Silicon Valley, astrobiologist Chris Chyba says, 'There's an argument that Mars was a safer place for early life than Earth.' Our planet probably had vast oceans since its youth, while Mars was a generally drier world. 'A sufficiently big impact on the early Earth would evaporate

Mars – although there's less transport of rocks this way, because they have to escape Earth's powerful gravity. And there was even a third player in the game.

'The numerical experiments,' Jakosky continues, 'suggest you could get rocks travelling from Mars to Earth or Venus, from Earth to Venus or Mars; and from Venus to Mars or Earth. So probably all three planets were exchanging rocks.'

Some of these interplanetary wanderers may have contained microscopic stowaways. Not just fossilised bacteria, like those in ALH 84001, but living microbes. Sheltered and cushioned inside the rock, living cells could survive the shock of blasting off a planet, the harsh radiation of space and the fiery descent to a new world.

'That opens the possibility,' enthuses Jakosky, 'that there has been exchange of life between the planets. You can't transfer bunny rabbits on meteorites, but microbes you could move back and forth. And that really does open up the question: on which planet did life originate?'

Wherever we asked, we found that one planet is becoming a clear winner.

On the leafy campus of Britain's Open University, Colin Pillinger – leader of a future Mars-lander mission – declares: 'Mars, being further from the Sun, may well have been conducive for life to develop earlier than Earth.'

Under the clear skies of Arizona, geologist Bill Hartmann contends: 'Liquid water and the conditions for life may have existed 50 million or 100 million years earlier on Mars than on Earth.'

the entire ocean, leading to an extremely powerful greenhouse effect,' Chyba continues. 'You'd bake the entire surface of the Earth to 1,500° for thousands of years, and sterilise the entire planet. If Mars never had large oceans, you couldn't create that intense kind of greenhouse, so it's harder to sterilise Mars with a big impact. In that sense, Mars would have been a better spot for the origins of life to happen than the early Earth.'

And that, in turn, leads to an idea that shakes us to the core. If life began on Mars, then the 'bugs' we are – possibly – seeing in the Martian meteorite ALH 84001 are not the desiccated remains of alien beings. They are in fact our own earliest ancestors.

'The hot thinking,' Chyba admits, 'is to say – gosh, life might have originated on Mars.' Conveyed to the early Earth inside their meteorite space capsules, the first bugs were destroyed by the hostile environment on the heavily bombarded Earth.

But one fine day a Martian meteorite broke open on the surface of an Earth that had calmed down from its tempestuous youth. The bugs spilled out on to a fertile new world where they could prosper, grow and evolve – through protozoa, fishes and reptiles to become mammals.

Bill Hartmann sums up: 'It's conceivable that the whole of life on Earth is actually Martian.' If so, that leads inexorably to one final conclusion, one that alters our perception of the place of human beings in the Universe. As astrobiologist Chris Chyba puts it: 'We may all be Martians.'

A MARTIAN CHRONICLE

CHAPTER 2

> "Ladies and gentlemen, I have a grave announcement to make. The strange object that fell at Grovers Mill, New Jersey, earlier this evening was not a meteorite. Incredible as it seems, it contained strange beings who are believed to be the vanguard of an army from the planet Mars."

Eight p.m., 30 October 1938, New York. A radio continuity announcer interrupts a CBS programme with the now-immortal lines: 'Ladies and gentlemen, I have a grave announcement to make. The strange object that fell at Grovers Mill, New Jersey, earlier this evening was not a meteorite. Incredible as it seems, it contained strange beings who are believed to be the vanguard of an army from the planet Mars.'

Orson Welles (right) and an actor in the CBS studios throw themselves wholeheartedly into the spoof 1938 *The War of the Worlds* broadcast.

The 'continuity announcer' was nothing of the kind. At just twenty-three years of age, Orson Welles was a brilliant actor and director (previously a bullfighter and magician) who was ideally qualified to undertake the task of dramatising H.G. Wells's novel, *The War of the Worlds*, for radio. He did it in a way that convinced and terrified the American nation into believing that they were under threat from an alien invasion.

The broadcast was an all-too-realistic stunt to improve audience ratings for CBS. Welles had been charged with putting on a series of Sunday evening radio plays for the station to compete with the far more popular *Charlie McCarthy Show* on a rival network. If he failed, his days in radio were numbered. Little did he know that on that particular night the gods were not just smiling on him – but grinning deliriously. Charlie McCarthy had an unknown singer on his show, and bored listeners started twiddling their dials...

Eight p.m. 30 October 2000, the Polecat Inn, Prestwood, Buckinghamshire. In an olde-worlde pub filled with stuffed animals and sober Home Counties residents having dinner, we meet up with Stephen Baxter – coincidentally, sixty-two years to the hour after the broadcast of *The War of the Worlds*. Baxter is young, perceptive, scientifically qualified, and now one of the hottest sci-fi writers in the world.

Amazingly, we all live within miles of each other, but this is the first time that Baxter and the pair of us have met. His telephone number is ex-directory, and we only succeed in tracking him down via an e-mail to Arthur C. Clarke in Sri Lanka, who acts as an intermediary.

'Orson Welles's *War of the Worlds* "faction" broadcast caused panic in the streets,' relates Baxter. 'Everybody was terrified. I'm sure it was because of the striking power of the image of an invasion from elsewhere – given that Mars was then a plausible base for life forms of some kind. Science fiction is always a metaphor for its time, so the original 1898 novel was really a parable about imperialism. In 1938 it was fears of the coming World War, transmuted into fears of the aliens.'

Welles certainly knew how to wind his listeners up. At one point in the broadcast, a 'reporter in the field' described an alien emerging from its spacecraft. 'There, I can see the thing's body. It glistens like wet leather. But that face. It... it's indescribable. The mouth is V-shaped, with saliva dripping from its rimless lips that seem to quiver and pulsate. The thing is rising up... the crowd falls back. I can't find words. I'm pulling this microphone with me as I talk. I'll have to stop the description until I've taken a new position. Hold on, will you please?... I'll be back in a minute.'

An 1898 illustration from the time of H.G. Wells's original novel, showing a shell destroying a Tripod – a Martian fighting machine.

A voice from Washington revealed that the Martians were landing all over the United States. Thousands of people had already been killed in cold blood by the aliens' death-ray guns. The broadcast culminated in the Martians sweeping through Manhattan and invading the radio station itself. It ended with a chilling, high-pitched scream.

The acting was so convincing that many Americans fled their homes before the play was over – which is when it became clear that the broadcast was a work of fiction. They hit the streets, hid in cellars, prayed, loaded up their guns and even wrapped wet towels around their heads as protection from Martian poison gas. Some even claimed to have *seen* Martians.

Welles was blissfully unaware of what he had precipitated until he bought a newspaper the

The acting was so convincing that many Americans fled their homes before the play was over

following morning. 'Radio listeners in panic,' screamed the headline. 'Many flee homes to escape gas raid from Mars.' The press were furious about the realism of Welles's presentation and strongly criticised him. People even brought lawsuits against CBS totalling $750,000, but all were eventually withdrawn.

However, there's no such thing as bad publicity. Welles's career took off instantly, and CBS was delighted to have America's most notorious actor broadcasting on its airwaves. In an ironic twist, Welles was on radio just three years later, on 9 December 1941 – the day of the Japanese attack on Pearl Harbor. When the (real) continuity announcer broke the news of the bombing, many Americans believed that they had been hoodwinked again!

The romantic myth: Mars and Venus enjoy a bath together in this fresco painted by Giulio Pippi.

It seems that Mars has always been associated with war in the collective public psyche. In 1000 BC the Chaldeans – inhabitants of what is now Iraq – called the planet Nergal after their great hero – the king of conflicts, master of battles, and the champion of gods. Similarly, the Greeks – whose civilisation flourished between 1200 and 100 BC – named the planet Ares, the son of the chief god Zeus and his wife Hera. According to Homer's *Iliad*, Ares was disliked by many of the gods – and the Greeks themselves – because he was tempestuous and murderous. But to the Romans, who glorified conflict, Mars was their revered god of war – and it was they who gave the planet the name we use today.

For nearly 2,000 years afterwards, Mars remained a red dot in the sky, slowly moving against the background of stars. It wasn't until Galileo pointed the first astronomical telescope to the heavens in 1609 that scientists realised that the planet had more to its character than mere symbolism. In 1659 the great Dutch astronomer Christiaan Huygens mapped a large dark spot on Mars – probably the region we now call Syrtis Major – and by logging its movement found that the Red Planet had a twenty-four-hour day. He also discovered the ice-covered southern polar cap, and – way in advance of his time – wrote *Cosmothereos*, a book on how planets might sustain life. It appeared posthumously in 1698 – in Latin – but by the end of the year it was available in English as *Celestial Worlds Discover'd: or, Conjectures Concerning the Inhabitants, Plants and Productions of the Worlds in the Planets.* Dutch, French, German and Russian editions appeared rapidly afterwards. It was the beginning of something very, very big...

So big that in present-day California (where else?), there is a whole research facility devoted to life on other worlds. The SETI Institute aims to flush it out, if it exists. The acronym derives from the 'Search for Extraterrestrial Intelligence', and, in an open-plan breezy barn of a building in the heart of Silicon Valley, perfectly sober scientists speculate on the nature of aliens and how we might make contact with them.

Frank Drake is the man behind this vision. Now chairman of the board of trustees of the Institute, Drake began his career in the late 1950s in a brand-new field of science: radio astronomy. Using a huge dish to tune in to radio waves coming from natural objects in space, like exploding stars and violent galaxies, Drake realised that his radio telescope could be used in reverse – as a transmitter. Radio waves travel virtually unimpeded through space, and the young Drake wondered if other life forms out there might be actively broadcasting to the Universe. Over the years he has assembled a team of motivated and multidimensional scientists around himself who 'listen in' to possible alien transmissions.

Forty years on, and the team still haven't had the phone call from ET. But on this beautiful sunny day at the height of California's summer, Drake doesn't look too despondent.

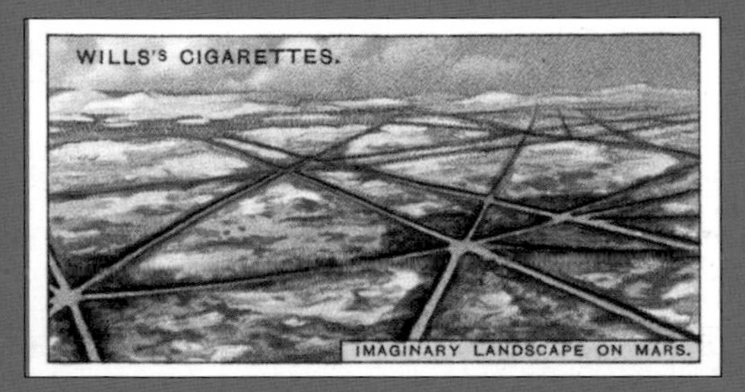

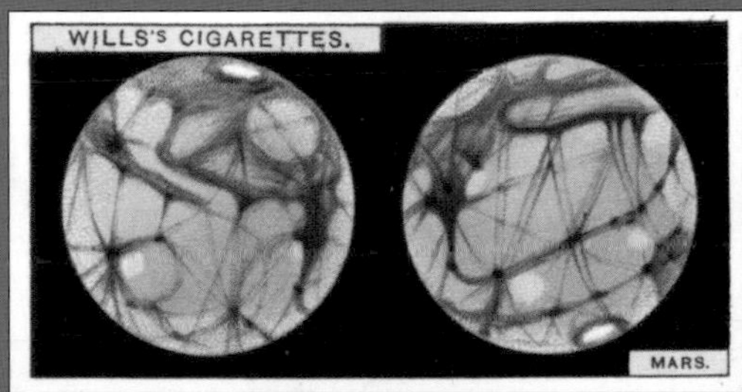

An instant source of knowledge on any topic, cigarette cards were included in packets in the early years of the twentieth century. This 1928 set shows Mars riddled with canals.

He muses on the whole question of whether life exists in the Universe – and in particular the possibility of life on Mars. 'The belief in life on Mars grew out of observations by an Italian astronomer, Schiaparelli, in the late nineteenth century. He believed he saw straight lines on the planet which he called *canali* in Italian – but, when read by English-speaking people, this was interpreted to mean "canals", which, of course, implies construction by a technological creature.'

In 1877 Mars and the Earth drew particularly close together in their paths around the Sun. Astronomer-historian Steve Dick takes up the story: 'Schiaparelli was observing out of Milan, but he didn't have a big telescope. He apparently did have very good eyesight, which is important. He did see some "canals" – but was very cautious in what he thought they might be.'

Frank Drake explains how the unfolding saga then leaped across the Atlantic. 'Percival Lowell was a banker and very rich. He became obsessed with the idea of life on Mars and used his fortune to construct the Lowell Observatory – the first high-altitude observatory in the world.'

Percival Lowell was an unlikely astronomer. He was born in 1855 into one of America's most illustrious families and brought up in the elegant Bostonian traditions of his day. It was said of the family: 'The Lowells speak only to the Cabots, and the Cabots speak only to God.' Lowell travelled the world extensively. He was refined, cultured and superbly well educated – fluent in French at the age of ten and Latin at eleven.

In his teens he took to astronomy. His brother Abbott Lawrence recalled: 'He read many books thereon, and had a telescope of his own of about two and a quarter inches in diameter. Later in his life, he remembered that with it he had seen the white snow cap on the pole of Mars crowning a globe spread with blue-green patches on an orange background.'

It was to be another twenty years before Lowell left business for astronomy. Clearly, his vision of Mars and what he knew about Schiaparelli's

In the heart of Silicon Valley, perfectly sober scientists speculate on the nature of aliens and how we might make contact with them

Percival Lowell, whose observations of the 'canals' of Mars – combined with his extensive writings on the nature of Martian culture – laid the groundwork for belief in alien life on the Red Planet.

discoveries never left him completely. The trigger was hearing that Schiaparelli's sight was failing. Lowell's life direction was now clear: he would assume Schiaparelli's mantle and find out the truth about the canals on Mars. In 1894, together with his friend Andrew Douglass (who later became famous for establishing the ring-counting system for dating trees), he founded an observatory over 7,000 feet up in the cool pine woods above Flagstaff in Arizona.

Today, the Lowell Observatory is famed as the place where Clyde Tombaugh discovered the planet Pluto in 1930. But in the 1890s Percival Lowell was using his extremely good telescopes to seek out signs of extraterrestrial intelligence instead. 'Lowell chose to interpret the canals as artificial constructions of the inhabitants of a dying planet who were trying to conserve their water,' explains Steve Dick. 'And there were tantalising hints that there was water on Mars, from the point of view of the polar caps.'

Lowell sketched hundreds of canals. What convinced him that they were artificial was their linearity, and the fact that they apparently changed with time – sometimes becoming double. 'If the planet possesses inhabitants, there is but one course open to them to support life. Irrigation must be the all-engrossing Martian pursuit... it must be the chief material concern of their lives,' he wrote.

'Irrigation must be the all-engrossing Martian pursuit... it must be the chief material concern of their lives'

He also mused about the nature of the intelligence that had constructed such an elaborate canal network. A committed pacifist, Lowell saw the global scheme as evidence that the Martians were united, and free from the scourge of war. Moreover, he believed that they were much more advanced than us. 'A mind of no mean order would seem to have presided over the system we see – a mind certainly of considerably more comprehensiveness than that which presides over the various departments of our own public works. Quite possibly, such Martian-folk are possessed of inventions of which we have not dreamed, and with them electrophones and kinetoscopes are things of a bygone past, preserved with veneration in museums as relics of the clumsy contrivances of the simple childhood of the race.'

Lowell was not one to keep his views to himself. As a result, what had been a demure astronomical debate about the nature of the canals turned into a raging public controversy. Even the severe economic depression of 1907 took its place on the back burner compared with the Martians. In that year, the usually conservative *Wall Street Journal* asked its readers: 'What has been in your opinion the most extraordinary event of the past twelve months?' For the *Journal* at least, it was 'not the financial panic which is occupying our minds to the exclusion of most other thoughts' but 'the proof afforded by astronomical observations... that conscious, intelligent human life exists upon the planet Mars'.

And it wasn't just Schiaparelli and Lowell who saw canals. 'Many other astronomers saw canals on Mars – maybe not as many as Lowell, and they certainly didn't have the same interpretation as he did,' explains Dick. 'The confusing thing was that, a lot of times, small telescopes could see them and big telescopes did not.'

The late Clyde Tombaugh – discoverer of Pluto and the most experienced user of Lowell's telescopes – thought he may have had the answer. We met up with him at Flagstaff fifteen years ago, when we were filming a nostalgic but incredibly thrilling re-enactment of his planet-find fifty-five years before.

Tombaugh related that, according to Lowell's own observation notes, he got the sharpest images of Mars when he 'stopped down' the telescope's lens with a diaphragm so that it measured only sixteen inches across. By doing so he managed to cut out the annoying haze of rainbow colours that plague all lens telescopes. But the bright, sharp images of the planet also have much more contrast – leading to the illusion of narrow, straight lines. 'I've seen some quite convincing "canal networks" in this way,' Tombaugh told us.

Now that spaceprobes have surveyed the surface of Mars even more thoroughly than that of Earth, we know that the canals don't exist. They were a by-product of astronomers straining their eyes through old-fashioned telescopes in days long before the electronic and digital revolution. Dick concludes: 'We now know that the canals are not there, so it was an optical illusion of some kind. It was a problem of the observations being at the very limits of the technology of the time.'

Canals or no canals, Lowell's passionate advocacy meant that belief in Martian life had become rife by the closing years of the nineteenth century. In the magnificent neo-classical library of the US Naval Observatory in Washington – built in the same era – Steve Dick talks us through some crazy but well-meaning plans to communicate with the denizens of our fellow planet. 'In the 1890s, when radio waves were discovered by Hertz, these "Hertzian Waves" were seized on by people like Nikola Tesla as being a good way for communication, not only on Earth but from Mars.'

A modern Tesla Coil undergoes a test. Capable of generating bolts of 'lightning' 13 feet long, generators like these were used by their inventor to communicate with 'Martians'.

'Oh, my,' laughs Frank Drake, 'I give a whole lecture on this. So here we go for forty-five minutes! The first experiment actually carried out was by Tesla in 1898, who got funding from J. Pierpont Morgan – the famous financier. He built an enormous radio transmitter – a Tesla Coil – on Colorado Springs. It had a tower 250 feet high, with a great ball on top. It would generate hundreds of thousands of volts, there'd be a great corona discharge from the top of the thing, and people's hair would stand on end for miles around. And Tesla, we don't know what signals he transmitted, but he claims to have received answers in the form of strange whistling, musical notes. We think he may have discovered "whistlers", which do indeed sound intelligent, but they're actually created by lightning bursts in the atmosphere of Earth.'

'The famous American astronomer David Todd proposed to go up in a balloon in 1909 with an early type of radio receiver,' relates Steve Dick. 'The idea was to get above some of the atmosphere so that more waves from Mars would penetrate. The scheme came to naught,

> **'Even though Lowell was wrong about the canals, he stimulated the imaginations of people'**

but in 1924 he came up with an even more interesting proposal.'

'David Todd proposed that all radio transmitters be turned off at the time of Earth's closest approach to Mars,' continues Frank Drake. 'And, indeed, many of the transmitters were turned off, but not all – and this resulted in several reports of signals being received. Some radio operators believed they received some Morse code from Mars, which seemed to spell out the word "zopp, z-o-o-p, zopp". It turned out that this was a word known to French people, and it was concluded very rapidly that this was a message in French from Mars. And that was totally believable, because the French were well aware that they were the most advanced civilisation on Earth, and therefore the Martians would surely emulate them.'

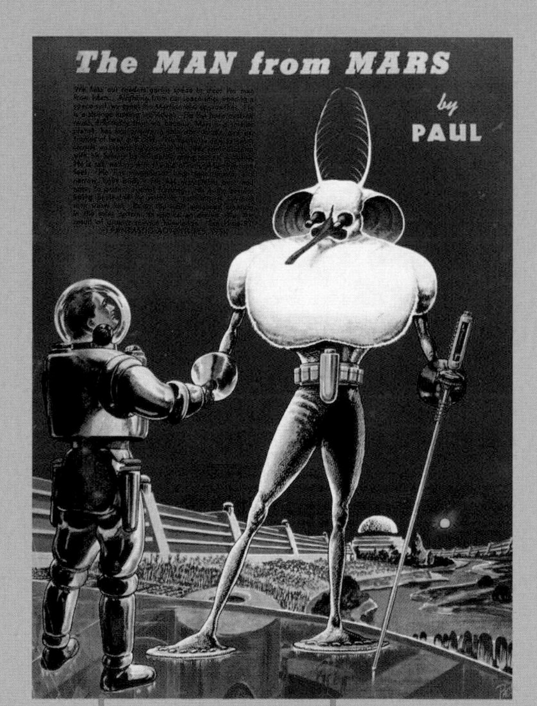

Man and Martian cordially shake tentacles, in an artwork by Frank R. Paul in the May 1939 edition of *Fantastic Adventures* magazine.

Even earlier – and even more bizarre – schemes have been mooted as a means of communicating with Martians. 'In the nineteenth century, and even in the eighteenth, you had people like William Herschel, the discoverer of Uranus, who believed that the Sun was inhabited,' reflects Steve Dick. 'It was thought that many of the planets of the Solar System were inhabited,' adds Frank Drake. 'One of the earliest proposals to communicate with life on Mars came from the famous German astronomer and mathematician Karl Friedrich Gauss, who suggested signalling to the Martians by planting a right-angled triangle – many miles across – made out of wheat and trees in the forests of Siberia. This would be visible through a good telescope on Mars, and would indicate that we understood the Pythagorean theorem.'

In the 1850s an ambitious scheme to communicate with Mars when at its closest was drawn up in Europe. As Frank Drake explains: 'It proposed using large mirrors which would be tilted so as to direct sunlight to Mars. These mirrors would be arranged in the shape of the Big Dipper – so that anybody looking at Earth from Mars would see an image of the Big Dipper flung across the continent of Europe. That one was never built.'

It seems that there's no end to humankind's ingenuity. Frank Drake: 'Von Littrow, who was a famous physicist, proposed constructing huge trenches in the form of geometrical figures in the Sahara Desert. These would then be filled with an inflammable liquid such as kerosene, and at night they would be set on fire, creating brilliant circles and triangles. This one was not funded until very recently, when Saddam Hussein carried out this experiment in Kuwait. Mars did not respond.'

'It's obviously a subject that lots of people took to heart,' observes Steve Dick. 'Even though Lowell was wrong about the canals, he stimulated the imaginations of people. It had a tremendous effect on popular culture, even in things like music – there was a Martian two-step! H.G. Wells knew about Lowell – who published his first book on Mars in 1895 – and followed with his own work, the novel *The War of the Worlds* in 1898. So that whole science fiction scenario, which began with H.G. Wells and has played down through the century – in both a positive and a negative way – stemmed, I think, in a large part from the imagination which flowed from Lowell's claim.'

Flash Gordon and friends enjoy a trip to Mars in 1938. Shame about the bad-hair day

Ask any sci-fi writer to name the most influential fictional book on Mars, and you get the same answer virtually every time. Arthur C. Clarke: 'Well, of course, *The War of the Worlds* is undoubtedly the best, and then *The Martian Chronicles*, and that's about it. I'm too modest to mention *The Sands of Mars*, of course, or *A Garden on Mars*.'

'H.G. Wells is maybe the first with *The War of the Worlds*,' pronounces Stephen Baxter. 'Edgar Rice Burroughs really fired the imagination of many readers, including Ray Bradbury, who wrote *The Martian Chronicles*. These were published after Arthur C. Clarke's *The Sands of Mars* – an early hard SF classic of the difficult life of the first pioneers on Mars. But pretty realistic spacesuits, and recycling the air – all that stuff.'

Each generation has its own Mars, both in science fiction and science fact

Why are Martians generally depicted as hostile? Despite Percival Lowell's belief in a race of pacifist and united canal-builders, H.G. Wells was probably to blame. 'Mars has always been a source of baddies,' observes Baxter. 'H.G. Wells's "vast, cool intelligences" in *The War of the Worlds* may be my favourites, just because of the opening sentence of that book: these creatures looking coolly, with longing, at our Earth... just a wonderful observation of that ancient but dying civilisation. It's a very modern novel, given that it's over a hundred years old, because it's a battle ecology. They were trying to take over our ecological niche because theirs had dried up and had become unsuitable for them. No compassion, no discussion – they just came and tried to take it. And I always enjoyed Gerry Anderson's Mars in the *Thunderbirds are Go* movie. First, explorers get to Mars and they find these great rock snakes that try to destroy us immediately.'

But Mars SF can have a more subtle message. In *A Martian Time Slip*, Philip K. Dick depicted Mars as a place of desolation and despair – a metaphor for the meaninglessness of modern life itself. Written in the early 1960s, it is a cosmic counterpart to Samuel Beckett's bleak worlds on stage, where life becomes an Endgame and Godot never comes.

But there's generally been a positive attitude about what Mars has to offer. 'It seems that when we look at Mars,' observes Baxter, 'the latest results always go for the most optimistic viewpoint. Let's face it, Mars is still the most hospitable place outside the Earth for mankind. I'm sure it's still going to be the main fount of dreams.'

Each generation has its own Mars, both in science fiction and science fact. 'The new generation,' Baxter opines, 'will be reading something like Paul McCauley's book *The Secret of Life*, about the Mars meteorite with the bugs. There's also a kind of nostalgic Mars – like Ray Bradbury's *Martian Chronicles*. And my book *Voyage* is about how NASA would have gone to Mars after Apollo, using kind of space shuttle technology, with swing-bys past Venus, nuclear-powered rockets and all that stuff. It never happened, and I think that there's always been a dichotomy about our view of Mars. It's the dream of the future, but also a nostalgic look back to the past.'

The images we project on to Mars fiction also reflect the political, sociological and ecological climate of the times. H.G. Wells not only saw *The War of the Worlds* as a parable of colonialism. He was obsessed by the evolution and future of mankind, and he saw in his Martians – with their huge brains – a possible future for us.

In the 1930s the influential American artist Isamu Noguchi conceived of using the Earth itself as a medium for sculptures on the cosmic scale.

His later commissions included gardens for UNESCO's headquarters in Paris and for IBM. But his greatest project – never realised – was conceived in the dark years following the Second World War. It was a giant human face, consisting of a vast set of earthworks, centred on a nose that was a pyramid a mile long. Fearing a nuclear war that would destroy humanity, Noguchi's face was designed to inform extraterrestrials that a civilised life form had once existed on Earth. In a strange mirror to the 'Face on Mars' that would inflame alien-culture freaks later in the century, Noguchi gave his 'Face on Earth' the title *Sculpture to be Seen from Mars*.

A few years later, Ray Bradbury was penning *The Martian Chronicles*. 'These books were written in the 1950s, after we knew that the planet could not be populated by golden-eyed humanoids and crystal cities,' says Baxter. 'Bradbury was writing about the loss of the past – the loss of the dream. You know – the destruction of Europe in the war. What a huge tragedy it must have seemed at the time to any sensitive person. And Stan Robinson's more recent Mars trilogy – about terraforming Mars into a world more like the Earth – is a metaphor for the green issues of our times.'

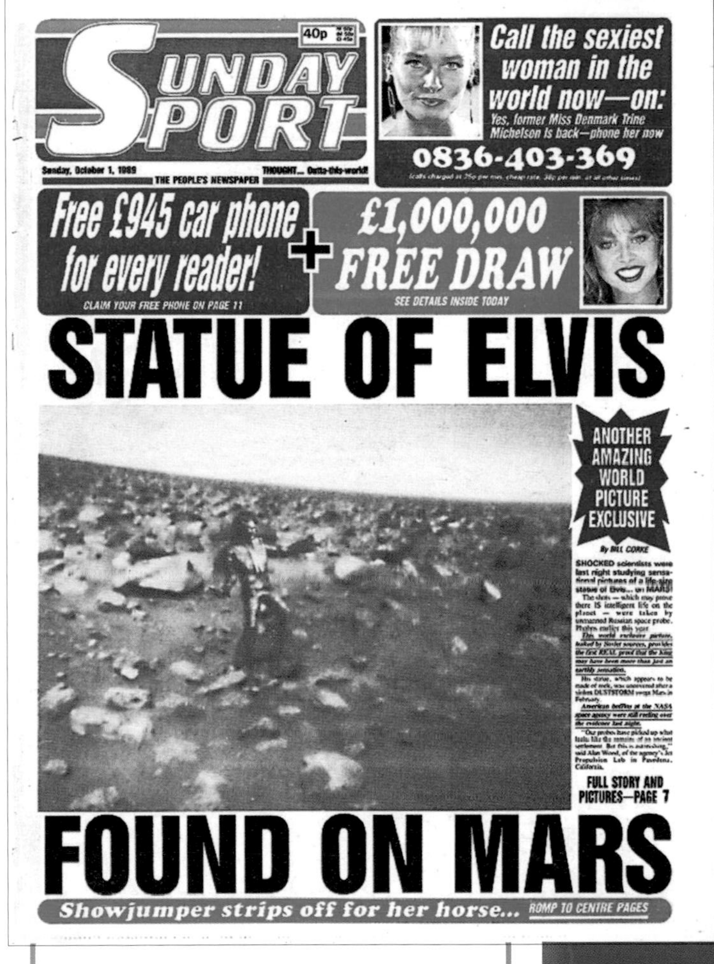

SUNDAY SPORT 40p

Sunday, October 1, 1989 — THE PEOPLE'S NEWSPAPER — THOUGHT... Outta-this-world!

Call the sexiest woman in the world now—on:
Yes, former Miss Denmark Trine Michelson is back—phone her now
0836-403-369

Free £945 car phone for every reader!
CLAIM YOUR FREE PHONE ON PAGE 11

£1,000,000 FREE DRAW
SEE DETAILS INSIDE TODAY

STATUE OF ELVIS

ANOTHER AMAZING WORLD PICTURE EXCLUSIVE

By BILL CORKE

SHOCKED scientists were last night studying sensa-[illegible] pictures of a [illegible] statue of Elvis... on MARS!

[illegible]

FULL STORY AND PICTURES—PAGE 7

FOUND ON MARS

Showjumper strips off for her horse... ROMP TO CENTRE PAGES

Even Elvis eventually ends up on Mars. One of the many famous *Sunday Sport* covers in the series that included 'World War II Bomber found on Moon' and 'Killer Plants stalk Queen Mum'.

Fiction it may be, but SF has had an enormously powerful effect on the science fact of the last fifty years – and on our vision of Mars. 'It was a huge pull on the imagination of the great space engineers – von Braun and those guys,' says Baxter. 'Von Braun was of course aiming throughout his career to get to Mars, and made a series of compromises in order to get to the Moon on the way. But the whole grand design, right from the Second World War days, was Mars.'

'We know how often the science fiction writers are correct,' adds Geoff Marcy, an astronomer who has discovered dozens of planets orbiting other stars. 'What we're seeing with the sociology of Mars is a bifurcation between what the public senses and what the scientists sense. You know, we scientists are so wrapped up in our equations and error bars that we suffer from the age-old syndrome of not being able to see the forest for the trees. We owe it to the public to provide information that is defended by the data, so we have to be conservative.

'But ultimately it is the questions a child would ask that we're trying to answer in science. And if you take your scientist's white lab coat off for just a moment, you realise that the most compelling destination for humanity is Mars – and that the most profound scientific questions reside on Mars and just beneath the surface of Mars.'

THE RED
PLANET

CHAPTER 3

‘Well, you’d wake up in the morning and it would be cold outside. And we’re talking extremely cold – colder than it ever gets on Earth. The sky would be different too. I’m looking out at the sky and I see clouds, but they’d be different and their colours would be different.’

The stars blaze in an uncannily black sky – hard untwinkling diamonds set in sable. The dome of night is not merely star-spangled. It is a blizzard of stars, where the familiar constellation shapes – Orion, the Plough – are lost among a myriad of fainter stars. Seeking a respite from the relentless throng of stars, your eye latches on to fuzzy nebulae, and on to dark interstellar clouds silhouetted against the sheets of background stars.

Dawn and dusk over the Martian plains are truly unearthly: dust in Mars's air tinges the sky pink and the low-lying Sun blue.

In the east, the first rose-fingered hints of dawn appear to lighten the sky. As light grows, the pink persists: there is no promise of a blue sky to come. In the strengthening light, a thin sheen of cloud glows in all the colours of the rainbow and all the lustre of mother-of-pearl.

Contrasting with the salmon-pink dawn, a brilliant blue Morning Star hangs above the eastern horizon. Look closely and you spot – in its dazzling rays – a duller 'star' close by.

A rugged landscape is appearing in the half-light. Jagged stones scattered across a desert landscape. Cliffs rising to high mesas. Curved sand-dunes marching off to the horizon. And everywhere, dust. Dust on the ground, dust on the rocks, dust smoothing every contour.

High above this alien panorama, something is catching the first rays of the Sun. It's a rough grey shape – a third the size of the Moon we're used to – and it's moving even as we watch. With its shape rapidly dwindling towards a thin crescent, this giant flying stone heads down towards the blue Sun as it rises through a thin reddish mist.

Welcome to Mars.

It's a world all too familiar to Richard Zurek, of NASA's Jet Propulsion Laboratory. He's been here many times – in imagination, at least. 'I got interested in planets in Junior High,' he begins, 'and Mars is always a fascinating place. It's Earth-like in some regards, and yet it's exotic enough that it has its own appeal.'

When we meet him, Zurek is not the happiest of scientists. He was in charge of the experiments on board NASA's two latest Mars missions – the infamous pair that failed to make it to the Red Planet in 1999. Models of the doomed Mars Climate Orbiter and Mars Polar Lander dominate Zurek's otherwise sparsely furnished office.

But transport him in imagination to the surface of Mars and present woes are forgotten. 'Well, you'd wake up in the morning and it would be cold outside. And we're talking extremely cold – colder than it ever gets on Earth. The sky would be different too. I'm looking out at the sky and I see clouds, but they'd be different and their colours would be different.'

He revels in the other sky-sights too, starting with that blue Morning Star. 'You'd be able to see lots of stars, but you'd see one new one that you wouldn't recognise from the Earth – because it *is* the Earth.' Its fainter companion is our planet's large Moon.

'On Mars,' Zurek continues, 'you wouldn't have the familiarity of the big Moon as we have on the Earth. The Martian moons are pretty small – in fact, if one of them were to pass in front of the Sun, instead of a total eclipse what you'd see is something a bit more like a doughnut, where an area of the Sun has been taken out, because the moons are too small to completely block the Sun's light.'

Zurek began his professional life studying the physics of the Earth's atmosphere. Now he's turned to the planet of his childhood fascination: 'In a way, I'm a weatherman for Mars.'

'While they might be a few hundred feet high on Earth, on Mars one of those dust devils might tower a mile or more in the Martian atmosphere'

So what's the forecast for a typical Martian day?

'You're probably going to see some early-morning clouds and some fog that dissipates during the day,' says Zurek. 'In the afternoon, you're probably going to see dust devils – the sort of things you might see kicked up by little whirlwinds here in the Mojave Desert. While they might be a few hundred feet high on Earth, on

Afternoon clouds hang over Mars's towering volcanoes in this view, which encompasses a quarter of the planet. These blue-white veils of ice crystals resemble Earth's feathery cirrus clouds.

Curved ridges and terraces of compacted ice and dust cover Mars's frozen north pole. Although the altitudes have been exaggerated in this view, the broad icy mountain still rises 8,000 feet at its centre.

The first post-card from Mars, sent back by Viking 1 in 1976. Familiar with the deserts of the American West, NASA scientists first adjusted the colour balance to show a blue sky.

Mars one of those dust devils might tower a mile or more in the Martian atmosphere.'

Frightening they may look, but fortunately these dust devils aren't as destructive as tornadoes at home. Mars is a smaller world than Earth, with a lower gravity, and the giant storms can grow under the influence of much lesser winds.

'If you were at higher latitudes, away from the equator,' Zurek continues his forecast, 'there are going to be some seasons where it'd be just like Wisconsin – cold fronts coming through, marked by clear areas and then cloudy areas. In fact, the storms on Mars seem to occur much more regularly than they do on the Earth, almost like clockwork every few days.'

But the Martian sky is really pink. It's coloured by fine dust particles whipped up from the desert, which has literally rusted away to ochre rocks and sand.

Come spring and early summer, and Mars may experience another planet-wide catastrophe never experienced on Earth: a global dust storm

Spring, summer, autumn, winter – Mars has seasons much like the Earth, except that each season is colder and more protracted: the Red Planet lies 52 per cent further from the Sun's heat, and a 'year' on slow-moving Mars lasts for almost two Earth-years.

Winter is a severe season indeed. With temperatures plummeting to –125°C, Mars's air begins to freeze solid. Even at the best of times, Mars has only a tenuous veil of an atmosphere of choking carbon dioxide gas – less than one-hundredth as dense as the air we breathe on Earth. At the winter pole, turned away from the Sun's heat, Mars's air freezes on to the ground as a brilliant white polar cap of 'dry ice'. In a bad winter, a quarter of Mars's atmosphere is frozen out.

Come spring and early summer, and Mars may experience another planet-wide catastrophe never experienced on Earth: a global dust storm.

Mars meteorologist Richard Zurek again: 'That's when a small dust storm or a cluster of small dust storms – for some reason we don't understand – just continues to grow, raising dust all around the planet. During the largest of these storms, you probably couldn't even see the Sun through the dust.'

Even a small dust storm would be no place to be, according to Mars geologist Bill Hartmann. 'The dust grains are picked up by the wind, like sand grains on a beach on Earth that hop along the surface and hit our knees. Well, on Mars, because of the lower gravity and the wind conditions, these can pick up to about two metres or so. On a bad sandstorm you could get into a kind of brown-out, be enveloped in blowing dust and not be able to find your way.'

The red desert dust hangs around for months in the planet's atmosphere, giving Mars's sky its unique pink colour. It's something NASA wasn't anticipating when the first spacecraft landed on the planet's surface. 'The first pictures released when the Viking landers touched down had a sky that was tinged blue,' Zurek recalls, 'because that was what we expected.'

Just as today you would adjust the brightness, contrast and colour balance of digital holiday snaps on your computer screen, so NASA scientists are used to tweaking the signals sent from distant spacecraft.

Olympus Mons – Mars's Mount Olympus – is the biggest volcano in the Solar System, rising to three times the height of Mount Everest. And it's probably still active...

'Then we realised by looking at the colour charts on the spacecraft that we didn't have the colour balance right,' Zurek smiles. 'When we adjusted it, there was a little bit of red to the sky.'

The colour of Mars's sky, the brown-out of dust storms and the Martian landscape are all matters of more than academic interest to Bill Hartmann. As well as working as a senior Martian geologist at the Planetary Research Institute in Tucson, Arizona, he spends several mornings a week in his artist's studio, re-creating the planets on canvas.

Hartmann was in the studio when we caught up with him. Since it was a phone interview, we had to request a verbal description of his work-in-hand. 'I have a Mars volcano painting under way, with a huge smoke plume. Looking round my studio, there's another that I was calling *Origins of Life on Mars*. I've just come back from Yellowstone Park, and I was looking at a hot spring and tried to paint this as early Mars, when life might have been forming.'

In all his Martian paintings, the colour of the sky is a crucial issue. 'We know that normally the sky has this pinkish tinge, but it's apparently very delicate. I've always wondered that if you looked upwards, that it doesn't perhaps get darker grey or blue, because of the very thin air.'

Astronomer-painter Bill Hartmann has a double international reputation – combined with the grace of modesty. It must surely be a pretty rare combination to be both an exceptional artist and a talented scientist?

'Well... I never thought about that very much in the first half of my career. My grandfather painted, and we used to entertain ourselves as kids. I did a bit of space drawing when I was a teenager, and then really put it all aside and went into physics.'

When Hartmann came to work on a textbook of astronomy, his background gave him a fresh perspective. 'A lot of textbooks in that period – the late 1960s – were pretty dry, and I thought that if I rendered some of these sights it would help students understand what these places in the Universe were like.'

In the past twenty years his perspective has broadened. 'I guess I approach things a bit more visually. I've always been more interested in that "first order of fact" – what would it be like to be there? – than, say, the fourth significant figure in an isotope ratio. And that's actually been very helpful because I've realised that most scientists are trained to go for the detail.'

An erupting Martian volcano, as depicted by astronomer-artist Bill Hartmann, who adds: 'This scene, I now feel, is possible on "modern Mars" (sometime in the last or next million years!).' Hartmann also suggests that the sky, after sunset, will turn from pink to blue.

Bill Hartmann was lured to physics and astronomy from an early age by the romance of the heavens. 'We had an encyclopaedia that had a map of the Moon with names on it. The names fascinated me; that there were actually places on the Moon, geographical places like missing kingdoms.'

The kingdom he currently surveys is Mars – in particular the huge volcanoes that rise from the Martian plains. Though Mars is only half the size of Earth, its greatest volcanic peak rises to three times the height of Mount Everest. This colossal mountain is wide enough to cover Spain. Appropriately enough, it is named Olympus Mons – Mount Olympus, home of the Greek gods.

'A lot of people say: "What would it be like to stand and see this huge volcano, Olympus Mons – the largest volcano in the Solar System – in the distance?" But it's really too big to see from the surface. It's like saying, "Let's paint England as seen from France".'

Olympus Mons is not a steeply rising peak, like Vesuvius. Its gradual slopes would hardly look like a mountain. To draw again on the Spanish analogy: imagine standing in Gibraltar and trying to spot a mountain – even one bigger than Everest – as far away as Madrid. Then imagine that it's not a steep mountain, but one that gradually rises up.

How does climbing Olympus Mons, then, rate as a challenge? We put the question to Steve Squyres from Cornell University, a Mars expert and also an active climber. 'Climbing the largest mountain in the Solar System would be kind of cool,' he agrees, 'but, you know, it would be pretty boring. Olympus Mons looks spectacular when you look down on it from orbit, but if you were standing on the surface it wouldn't be much of a climb – it'd be a drudge.'

Though Mars is only half the size of Earth, its greatest volcanic peak rises to three times the height of Mount Everest

In the pictures sent back by orbiting spacecraft, Olympus Mons appears as a massive and inert beast. For decades, astronomers have thought that the Martian volcanoes are extinct. But Bill Hartmann now has evidence that Olympus Mons and its kin are only lying dormant. In the future, Mars will again live up to his canvas of the volcano erupting smoke.

'The Mars I'm studying today is not the Mars that I was studying five or ten years ago'

'What we've been doing is looking for the youngest lava flows, the ones with fresh textures and not very many impact craters,' Hartmann explains. Rocks from space are constantly falling to Mars and blasting out craters. From the moment a lava flow solidifies, it starts to bear these scars. So the more craters, the older the lava flow; the fewer craters, the more recently the lava flow has erupted.

'When we looked for the youngest ones, we found flows that come out less than 100 million years old – maybe even down to 10 million years,' he continues. 'Now that sounds like a long time on the human scale, but this is getting within the last 1 per cent of the history of the planet.'

And there's more evidence too. In Boulder, Colorado, Bruce Jakosky invokes the evidence from Martian meteorites that have been found on Earth. 'The youngest is just a couple of hundred million years old,' he says, 'so Mars was active as recently as that.' These meteorites have been blasted out of just a handful of random locations on Mars; if one of them is as young as 200 million years, then the law of averages says there must be other lava flows even younger.

'These two facts – completely different ways of getting the age – are in beautiful agreement,' Hartmann enthuses. 'And it's unlikely this volcanism continued up to the last 1 per cent of the history of Mars and then shut off. So the bottom-line inference would be that – yes – volcanism is undoubtedly still going on on Mars. And there will be volcanoes erupting in the future.'

Jakosky concurs: 'I don't think anyone doubts that there's active volcanism somewhere on Mars at a low level today.' And that's not what you'll read in the standard textbooks, which typically state: 'Mars has volcanoes that are inactive today.'

It's part of a new and invigorating perception of our neighbour world. No longer is Mars seen as a dead and run-down planet, where everything of interest has happened in the past. As Jakosky puts it: 'The Mars I'm studying today is not the Mars that I was studying five or ten years ago.'

Fellow Mars geologist Mike Malin puts it in a historical context. 'In the 1960s, Mariner 4 and Mariners 6 and 7 showed us a cratered planet, with enough atmosphere to blow the dust around. Mars was then considered as "the Moon with an atmosphere".'

Later spacecraft revealed the huge volcanoes, giant canyons big enough to swallow a mountain chain as big as the Alps and swarms of dry valleys – apparently the tracks of past rivers. Fuelled by visionaries such as Carl Sagan, people began to expect to find life on Mars when the Viking missions landed in the 1970s. 'At that time, people began to think of Mars as "the Earth with craters",' Malin continues.

'But, in fact, Mars is neither the Earth with craters nor the Moon with an atmosphere,' Malin emphasises. 'Mars is Mars. Mars is its own unique planet that has its own unique way of telling us things about itself.'

Malin has put Mars's landscapes under the microscope, with images from his high-magnification camera aboard the orbiting Mars Global Surveyor spacecraft. The camera can

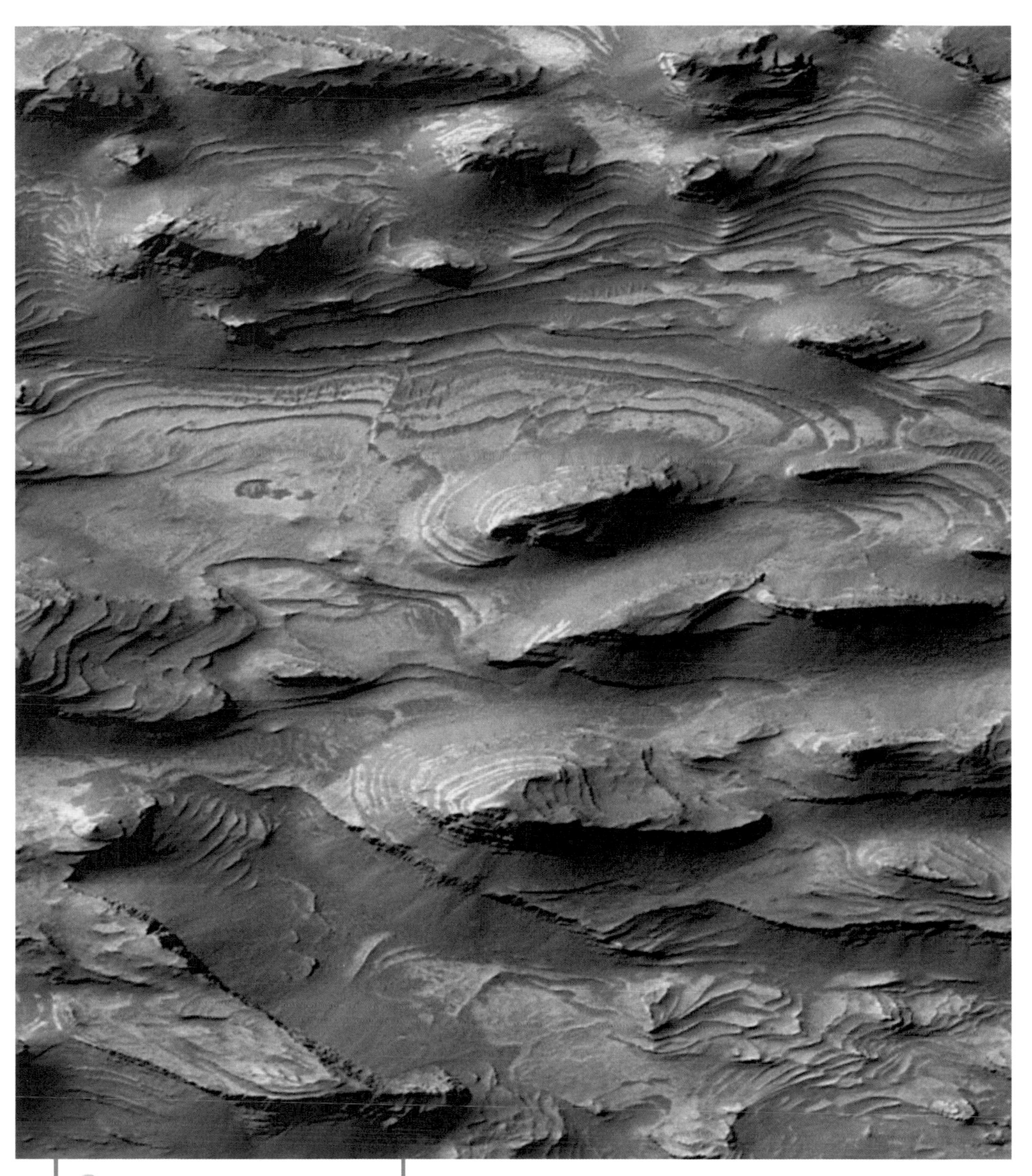

The floor of Mars's giant canyon system resembles a layer-cake in this recent image. The origin of this layering is still a mystery: recurrent giant dust storms, eruptions of volcanic ash or dust, or deposits in a vast dried up lake.

Mars, warts and all! Not a photograph, but a globe created from precise measurements of heights on the Red Planet, with the altitudes exaggerated. Far left is the monstrous volcano Olympus Mons, with the immense canyon system of Valles Marineris across the centre.

reveal details on the planet's surface as small as an estate car. Under its intense scrutiny, Malin can see Mars changing.

'We're looking at dust slides on escarpments say a kilometre or two long, and you might see a dust slide occur on that slope about once a Martian year. So it's not as dynamic an environment as snow in a mountain range on Earth, where you have dozens of avalanches; nor is it as slow as the rock avalanches on the same slopes which occur every century or so.'

In fact, Malin's team is finding it increasingly difficult to explain Mars in terms of what geologists know about the Earth. 'It's very hard to work out what's going on when all you have are pictures,' he admits, 'because we're not seeing the geological process – we're seeing the handiwork of the process. So it's seeing skid-marks on the road and inferring that a car had to brake.'

That's an easy one in comparison with what the Mars Global Surveyor is revealing. 'When we're seeing features on Mars, we are limited by our experience of terrestrial processes,' Malin continues. And with a nod towards his younger colleague, Ken Edgett, Malin concludes: 'And then he points out, "Hey, this isn't Earth, so why are we using terrestrial processes?"'

'We keep trying to use terrestrial analogues,' Edgett explains, 'and we've come up short. We can look at Mars, and we can say: "Well, here we have sand-dunes, dust devils, frost, ice." But then you quickly run out of analogues. On Earth, rainfall and running water do a lot to shape our landforms, and it doesn't rain on Mars right now.'

Water. That's the one word that sums up the paradoxes of Mars; the one substance that may hold the key to understanding this unique planet.

Ever since Percival Lowell proposed that intelligent Martians had built canals to transport water from the frozen poles to the equatorial deserts, astronomers have pondered the role of water on Mars. The space missions of the past forty years have destroyed the myth of the canals, but have only deepened the mysteries of Mars.

Water. That's the one word that sums up the paradoxes of Mars; the one substance that may hold the key to understanding this unique planet

'There's no liquid water at the surface of Mars,' says Bruce Murray, who was in charge of the Viking missions of the 1970s. 'It's just too cold at the surface – it's like Antarctica.'

Yet pictures from the same missions showed clearly the handiwork of flowing water: vast gouges through the landscape, apparently carved by enormous flash floods, and also much smaller winding channels, looking for all the world like dried-up river systems. They are silent witnesses of immense geological forces in Mars's distant past.

To help us unravel the history of Mars, we travel to Menlo Park, in California's Silicon Valley. In the quiet suburbs lies a campus of the US Geological Survey – built right on top of the pent-up energy of the San Andreas Fault. We're welcomed by Mike Carr, an English geologist who has transferred his allegiance twice over – to the US and to the planet Mars.

'Mars is a geologist's paradise,' he enthuses. And – echoing Ken Edgett's words – he adds: 'And it's also always providing surprises. It seems we just start beginning to understand what's going on, then – boom – something comes along and it just blows us out of the water, and we have to start again.'

In the beginning, Mars – like the Earth – accumulated from rubble orbiting the youthful Sun. The last pieces of rock to fall in from space smashed out craters in the planet's surface, which we still see strewn across the highlands of southern Mars.

'Then the planet started to experience almost all the geological processes we see on Earth,' says Carr. 'But for some reason the results were much more dramatic.'

'Come and take a look at this.' He leads us from his office to the adjacent large workroom. From an earlier career in geophysics, we recognise the large tables where researchers pore over maps, geological charts and aerial photographs of the Earth.

But on display here is a 3-D view of a world that is clearly not our own. Half is peppered with craters, dominated by two enormous indentations in the planet's crust. Vast cones rise from its surface; enormous cracks spread a quarter of the way around the planet; and huge tracts of the surface look as though they've been washed away.

'It's not a photograph,' Carr adds. 'This image has been built up from altitude measurements from the Mars Global Surveyor.' NASA's latest Mars mission has surveyed the Red Planet to a greater precision than we know the Earth's surface.

Carr picks out the cones: 'We have these huge volcanoes, much bigger than anything on Earth,' he begins. Volcanoes on our planet are continually kept down to size by the Earth's drifting continents – the inexorable motion of the planet's surface known to geologists as 'plate tectonics'. Molten rock punches its way up through the Earth's solid crust and starts to build a monstrous steaming mountain; but before it can grow too big the crust has moved on. The stream of lava must start again, punch a new hole, and build a fresh volcano.

'On Mars there's very little indication of plate tectonics,' Carr concludes, 'at least throughout most of the planet's history.' As a result, an up-welling stream of lava builds a single volcano that grows ever higher and ever wider.

Mars – the dynamic planet. Landslips run all along the edge of Mars's huge canyons, signs of continual erosion. And the small cloud (bottom left, casting a shadow) is a dust devil – a tornado lifting dust from the deserts.

'There's a vast canyon system – we still don't know exactly how it formed. You could drop the Grand Canyon in it and it would just get lost'

Now he fingers the great cracks on Mars. 'There's a vast canyon system – we still don't know exactly how it formed. You could drop the Grand Canyon in it and it would just get lost.'

The details may be obscure, but the great canyon of Valles Marineris seems to hint at frustrated geological forces. Thanks to plate tectonics on our planet, the San Andreas Fault beneath our feet will eventually give way and release its geological tensions in an earthquake. Mars – with its immobile crust – may just split open.

'Then we have these enormous floods,' Carr continues, pointing out the regions of Mars's surface that seem to have been washed away wholesale. 'That's evidence of past climatic change. The present climate is very hostile, but there are these indications of different conditions in the past.'

Carr goes on to outline his conception of a warmer, wetter period in Mars's remote past. 'I think that very early Mars had a period when liquid water was stable on the surface, and we had rain and lakes and oceans and all those kinds of things.'

But how much water? The Mars community is deeply divided. Jim Head has studied all the rocky planets, and he's at the truly watery end of the spectrum of opinion. At Brown University in Rhode Island, Head has probed the vast flat northern plains of Mars.

'Several years ago, it was hypothesised that there was once an ocean here,' says Head, 'but the idea was very controversial. I've always been wary of the idea of large-scale standing bodies of water on Mars. With the Global Surveyor altimetry data, we are now in a position to test the oceans hypothesis.'

As the orbiter swung around Mars, it crossed the boundary between the northern lowlands and the surrounding highlands more than a thousand times. Head has discovered that it's at exactly the same altitude all the way round, just what you'd expect for the shoreline of an ocean. And the plains are exceptionally smooth, flatter than anything on Earth apart from dried-up lake beds. Head insists: 'That's consistent with the surface having been formed by sedimentation at the floor of an ocean.'

In a computer model, Head has then 'refilled' the northern lowlands with water. It takes the same volume of water as we find in the Earth's Arctic Ocean. And, like the Arctic Ocean, Mars's polar seas would undoubtedly have been covered with a thick layer of ice.

But not everyone is convinced. Over in Boulder, Colorado, Bruce Jakosky quips: 'I don't think the evidence really holds water!' He points out that there are two supposed boundaries around the northern plains, at different levels. The outer one is clearly not at the same level all the way round, and no one is now invoking an ocean to explain it. 'Even the inner one is not quite a perfect bathtub ring – so why not say it's due to whatever process produced the outer ring?'

Jakosky does believe that water flowed on the young planet, but on a more modest scale. 'On the oldest surfaces,' he points out, 'you see patterns that look like branching weather systems, tree-like patterns of valleys. It's been widely accepted for a couple of decades that they were formed by liquid water flowing over Mars's surface.'

This idea faces one major hurdle. If water was once flowing in these now-dry valleys, then Mars must have been a lot warmer than today's subzero temperatures. That means the planet was coddled in a much thicker atmosphere, creating a beneficial greenhouse effect. So where has all that atmosphere now gone? Theories abound: it was blasted into space by meteorites, stripped away by particles from the Sun or absorbed into Mars's surface.

As we probed more and more Mars experts, though, we found that not everyone now believes early Mars was a warm Eden. Mars meteorologist Richard Zurek thinks there's a simpler answer. 'We know that early Mars was wetter – we can see the evidence that water flowed across its surface. Whether it was *warmer* is an open question.'

Zurek adds: 'There's some speculation that we really don't need a warm atmosphere for liquid water to have flowed for some distance across the planet.' He envisions the heat of subterranean lava melting a patch of ice, to release a stream of water that flowed over a frozen planet.

Pascal Lee agrees. Based at NASA's Ames Research Center in Silicon Valley, he's forsaken his Californian summers for the chill of the Arctic. Lee believes that the clues to Mars will not be found in the hot deserts familiar to most NASA researchers, but near the barren, frozen poles.

'Here we have very strange networks of valleys and canyons, which look astonishingly like the ones on Mars,' he says. 'We feel there's a common history, and it all ties in to a very cold past on Mars – not an early Mars that would have been warm and wet.'

In this new view, early Mars was wet, but it was generally frozen all along. 'The valleys and canyons would have formed in association with ice cover, ice sheets and glaciers – not so much the ice itself ploughing through the valleys but often-times by the ice melting.'

And that raises the most profound question concerning the Red Planet's geology. If Mars was scoured by water in its frozen past, then can we expect to find water in its frozen present?

Until June 2000 few scientists would have been quoted as saying 'yes' – so deeply rooted was the idea of Mars as a world that has now frozen solid. But a major press conference at NASA destroyed any remaining complacency that we understand the Red Planet.

On the podium were Mike Malin and Ken Edgett, with the latest from their 'microscope on Mars', the camera aboard the Mars Global Surveyor. Knowing the sensational interpretation they were about to present, NASA also invited veteran Mars geologists Bruce Jakosky and Mike Carr to comment.

Malin kicked off by showing gullies flowing down steep slopes on Mars – on valley walls and the sides of craters. 'Had this been seen on Earth, there would be no question water is associated with it,' Malin asserted. So far, so good: the gullies were a small-scale addition to the catalogue of dry valleys and flood-scoured plains.

Then Ken Edgett dropped the bombshell. He reported that the gullies had washed down rocky debris, which had spread out to form a wide apron. Examining these aprons with his geologist's eye, Edgett concluded that they must be 'very, very young – no more than 1 or 2 million years old'. On the geological timescale, liquid water has flowed on Mars within the last twinkle of an eye.

'I had to be dragged kicking and screaming to this conclusion,' Edgett recalls, as he was forced to throw out all accepted wisdom about water on Mars. After discovering the gullies, he and Malin set out to date their rocky aprons by counting craters on them – just as Bill Hartmann had previously worked out the age of lava flows on Mars's volcanoes. Only there was a problem. 'When we looked,' Malin told us later, 'we could find no craters on them.'

The aprons are so recent that not one of them had yet experienced the random strike of a meteorite. In fact, with no craters to count, Malin and Edgett can't actually date the aprons, can't say how long it's been since the gullies were flowing. 'They are perhaps a hundred times younger than Bill Hartmann's lava flows; in one sense they are literally immeasurably younger, because we can't count any craters on them.'

Pushed for a figure, Malin says: 'We could be talking about a couple of million years, or it could be right now.'

[Edgett] was forced to throw out all accepted wisdom about water on Mars

Though they don't know exactly when each gully last flowed, statistics tell us that some must be flowing at the present day. Like Hartmann's lava flows, it's inconceivable that gullies scattered all over the planet were active until a million years ago and have now all switched off in unison.

The discovery threw Mars researchers into turmoil. Mike Carr admits he was left floundering. 'There's no question that those things look like water-worn valleys,' he admits when we catch up with him a few weeks later. 'But present-day Mars is so cold – 100°C below freezing – how in the world would you get this liquid water coming out of the ground?

'At the press conference, I suggested liquid carbon dioxide instead,' Carr continues. He thinks a pool of Mars's atmospheric gas might be trapped under ground, under pressure, until it's released by a landslide: 'And then we have liquid carbon dioxide flying out. That sounds very exotic, I know, but this is an exotic planet.'

Carr is clearly agitated as he speaks. Two sides of the scientist are competing in his mind. 'It's a question of do you believe the physics or do you

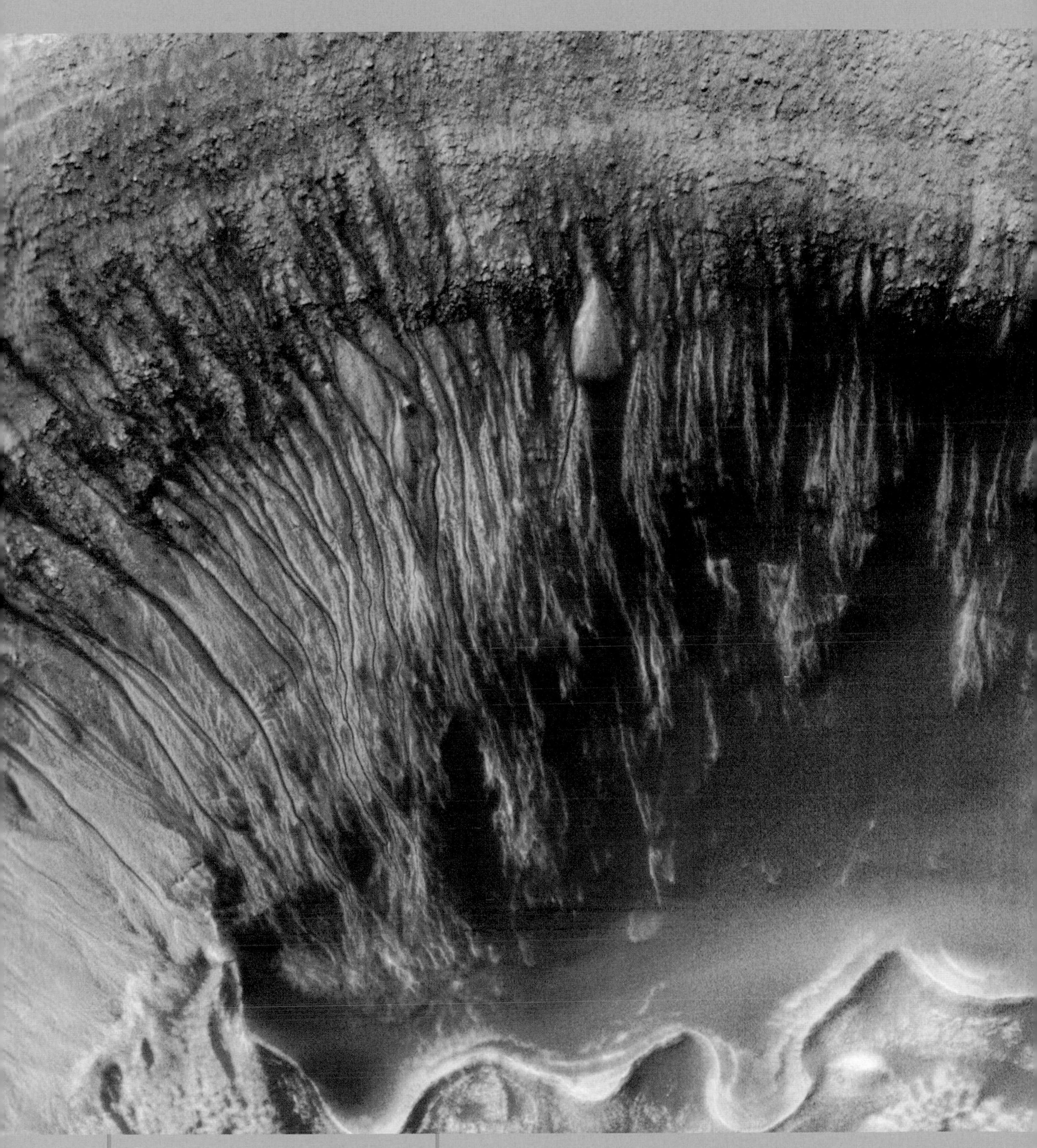

Springs of liquid water have apparently gushed from the walls of this Martian crater in the very recent past. These water-worn gullies – on a planet that's permanently frozen – are giving scientists a major headache!

believe the geology? The physics says, "It's impossible for it to be liquid water", but the geology says, "My God, it's liquid water". So it's a tough call.'

'The physics says ,"It's impossible for it to be liquid water", but the geology says, "My God, it's liquid water". So it's a tough call'

Bruce Jakosky recalls Carr's dilemma. 'When you press him, he'll say he thinks it's liquid water – but he's so confused by that that he wants to explore alternatives.' Jakosky himself has no doubts. 'It's gone beyond questioning whether it's indicative of water, to talking about the mechanism for getting liquid water there. I've heard three or four mechanisms proposed, none of which I really believe – not even my own!'

Frost begins to glisten on the floor of Martian craters as winter approaches. This giant crater is called Lowell, after the astronomer who devoted his life to the Martian 'canals'.

The nearest analogy to these gullies on Earth is a natural spring. In Mars's cold climate, though, the underground water should freeze solid – and if any should reach a Martian spring, the water should evaporate straight into the thin atmosphere.

At the press conference, Malin and Edgett suggested that a plug of ice builds up at the spring line, confining the underground water under pressure. When the ice plug fails, a stream of water gushes down the slope before it evaporates away.

But how come the water in the soil isn't frozen solid? Volcano expert Bill Hartmann thinks it's warmed by heat coming from within the planet. 'It's like Yellowstone Park some place on Mars, and it's melting the bottom side of the permafrost ice layer and creating an aquifer of liquid water. So you've got underground rivers flowing around Mars, and every once in a while these come to the side of a crater wall or hillside and burst out of the side of the hills.'

Jakosky's mechanism elaborates this idea. 'I mentioned this at the press conference,' he says later, 'and it probably didn't occur to Malin and Edgett since they live in San Diego and before that in Phoenix – both warm-weather climates. Here in Colorado it gets cold enough in winter for the pipes to freeze. If you want to keep the pipes from freezing solid, you just turn on the faucet a little bit. And basically the water goes through the pipe faster than it can freeze.'

Water is welling up from Mars's warm interior, in Jakosky's model. It flows through the chilly surface layers 'faster than it can freeze' and emerges in the gullies through a naturally slowly leaking tap.

But here's where Jakosky has to pull the plug on his own idea – and Hartmann's. 'The fatal problem with this mechanism is: what causes the water to come up? One of the drivers on Earth pushing water up is hot underground magma, like at Yellowstone. But you don't see recent volcanic activity connected to these break-outs on Mars, so it's unlikely that would work.'

If you can't invoke volcanic heat, you might expect to find springs of liquid water where it's naturally warmest – in suntraps around Mars's equator. But the exotic planet has more surprises in store. Most of the gullies lie far from the equator, and face away from the Sun's warmth.

'That leads to an explanation that invokes the changing tilt of Mars's axis,' Jakosky continues. As an old-fashioned school globe shows, the Earth's axis is tipped at an angle of about 23° to its orbit around the Sun; and Mars today has a similar tilt. 'But Mars's tilt varies dramatically with time,' explains Jakosky, 'primarily due to the gravitational tug from Jupiter.'

Mars can tip so far over that summer sunshine can fall directly on the slopes where the gullies are seen. Here, the temperature could have risen high enough to melt the ice down to a depth of perhaps a thousand feet, and springs would gush forth.

Polar expert Pascal Lee also invokes Mars's changing tilt. 'These oscillations occur on timescales of about 100,000 years, which is very recent by the standard of Mars's history, and the climate can certainly change quite rapidly over that time.' But his Arctic experience leads to a different explanation for the gullies, invoking not gushing springs but melting patches of snow.

'When the planet is tilted to maximum,' he continues, 'the polar caps might evaporate dramatically and ice would be redistributed around the planet. It wouldn't be surprising if transient snow patches were to occur here and there at the latitudes where Malin and Edgett have reported their features.'

Lee has investigated nooks and crannies in the Earth's Arctic and Alpine environments and has discovered that 'the repeated melting of these localised snow patches results in gullies that have exactly the same morphology as the ones that have been recorded on Mars'.

But perhaps we don't even need to find a way of raising the temperature above freezing point. David Wynn-Williams is a British researcher specialising in another frozen waste – Antarctica. 'Don Juan Pond, in one of the Antarctic dry valleys, is at –50°C, and it never freezes.'

The secret is natural antifreeze, as Nathalie Cabrol – a vibrant young French researcher at the Ames Research Center – explains to us. 'You can have liquid water at those latitudes if it's a super-brine,' she enthuses. 'That's water loaded with salts of potassium, magnesium and so on. You'd have a lot of work to do before you drink that water if you go to Mars! With a super-brine – a hypersaline solution – you can depress the freezing point quite low: you can get to between –20° and –60°.'

Wynn-Williams has explored the possible geology of super-brines on Mars. 'At the edges of the poles,' he explains, 'the pressure of the ice above pushes down on these hypersaline solutions, squeezing them horizontally beneath the surface until they drain down into the gullies. They stay liquid long enough actually to create the channels before they evaporate and produce the fans or aprons – deposits of salts.'

Cabrol goes even further. When we ask her, 'Do you believe there's still water on Mars?', she surprises us with a forthright, 'Yes I do.' So – tell us more...

When we ask her, 'Do you believe there's still water on Mars?', [Cabrol] surprises us with a forthright, 'Yes I do'

'In the past few years, I've been studying evidence for past lakes inside impact craters,' she begins. Cabrol has followed Mars's now-dry valleys downstream and discovered many cases where a valley ends up inside a large old crater. Within the crater, she's found sediments that resemble the dried-up bed of a lake – lacustrine deposits.

'While we were doing this,' Cabrol continues, 'we wanted to know how old were those lakes.' Like Hartmann with his volcanoes, and Malin and Edgett with their outflow aprons, Cabrol turned to the tried and tested method of dating a planetary surface – counting the small craters accumulated as meteorites have fallen to Mars. 'Many of them were quite old, but on some of

The ups and downs of Mars are accentuated by false colours in this computer-generated view. The southern hemisphere (foreground, red) is a vast upland region, with giant volcanoes rearing upwards (left) and the huge canyon system cutting through it (right). The flat lowlands in the northern hemisphere (background, blue/green) may be the floor of an ancient ocean.

these lacustrine deposits you couldn't see one impact crater superimposed.

'So that piqued our curiosity,' says Cabrol. 'They are young, but how young I don't know. They might be 200 million years old; they might be yesterday's lakes. And then something really incredible happened. But this is why science is fun.'

Cabrol began to dream of her ideal geological mission to Mars, and turned to her colleague Bob Haberle for advice. Bob – like Richard Zurek – is a weatherman for other worlds, and he'd already mapped out where he'd like to establish a network of meteorological stations on Mars. They were places where the Sun's heat might just melt ice into water on the hottest of summer days.

'I had just finished my mapping of the lakes of Mars,' Cabrol beams, 'and I went back to my office for the map. We found that my most recent lakes exactly matched Bob's regions!'

Surely it could be no coincidence that the regions where liquid water *could* exist on Mars are exactly the places where Cabrol has found geological evidence for lake deposits that appear totally fresh. 'That's why we believe there is water on Mars today.'

'We're not the only people talking about recent to contemporary activity,' she points out. 'Bill Hartmann is seeing very young volcanic features, while we are seeing very young lake features. And there's Malin and Edgett with their seepage features. So the whole picture of Mars is shifting from a dead planet to something that is certainly active today – how active, we don't know.'

And we'll let Ken Edgett – co-architect of the new revolution – have the final word on the Red Planet as viewed at the turn of the millennium. 'It's very clear that the Mars we're now seeing – every crater, every trough, every valley, every layer in the valley walls – speaks to us, and tells us that Mars is not what we thought. It also tells us that Mars isn't going to give up its secrets easily...'

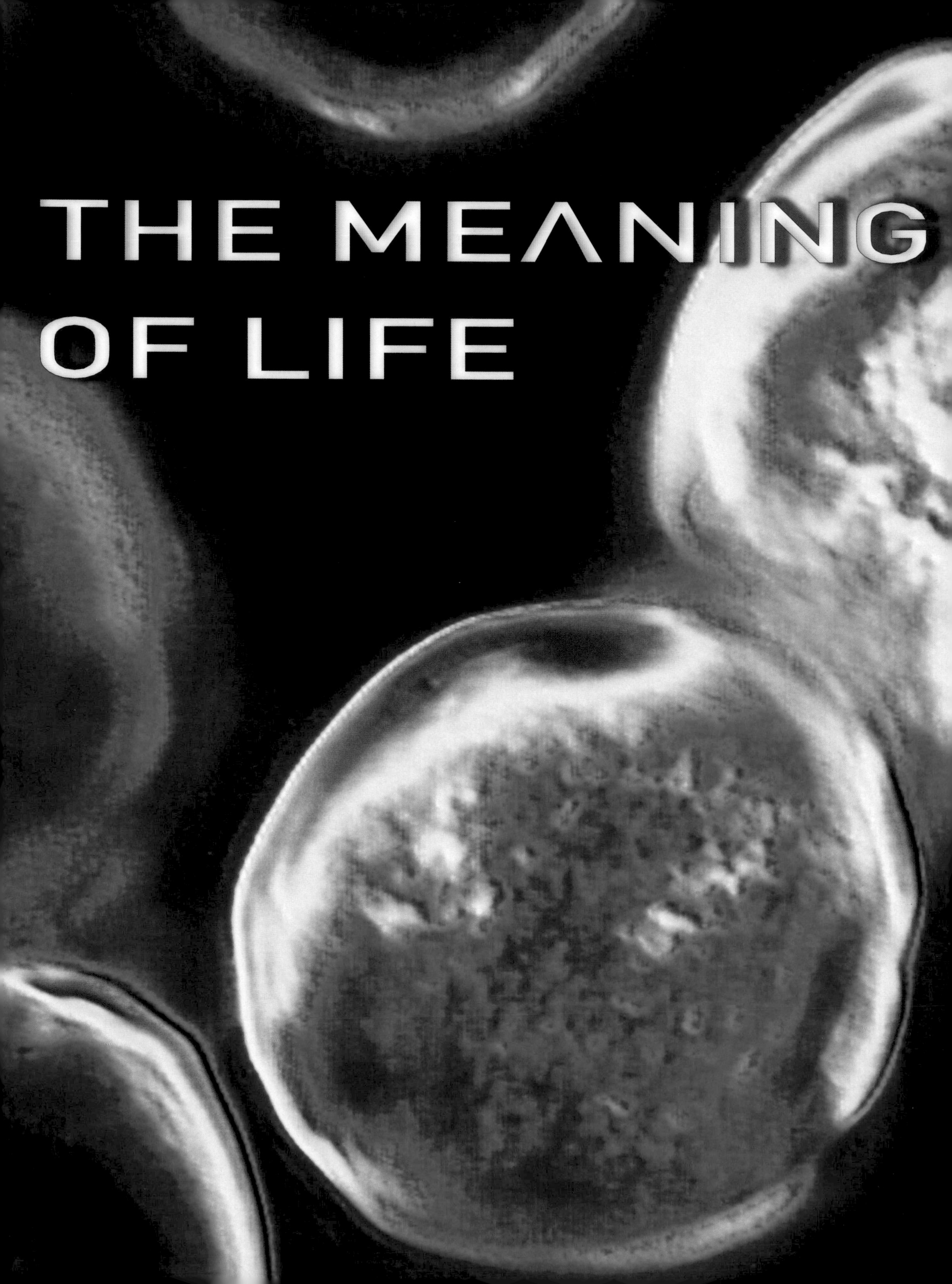
THE MEANING
OF LIFE

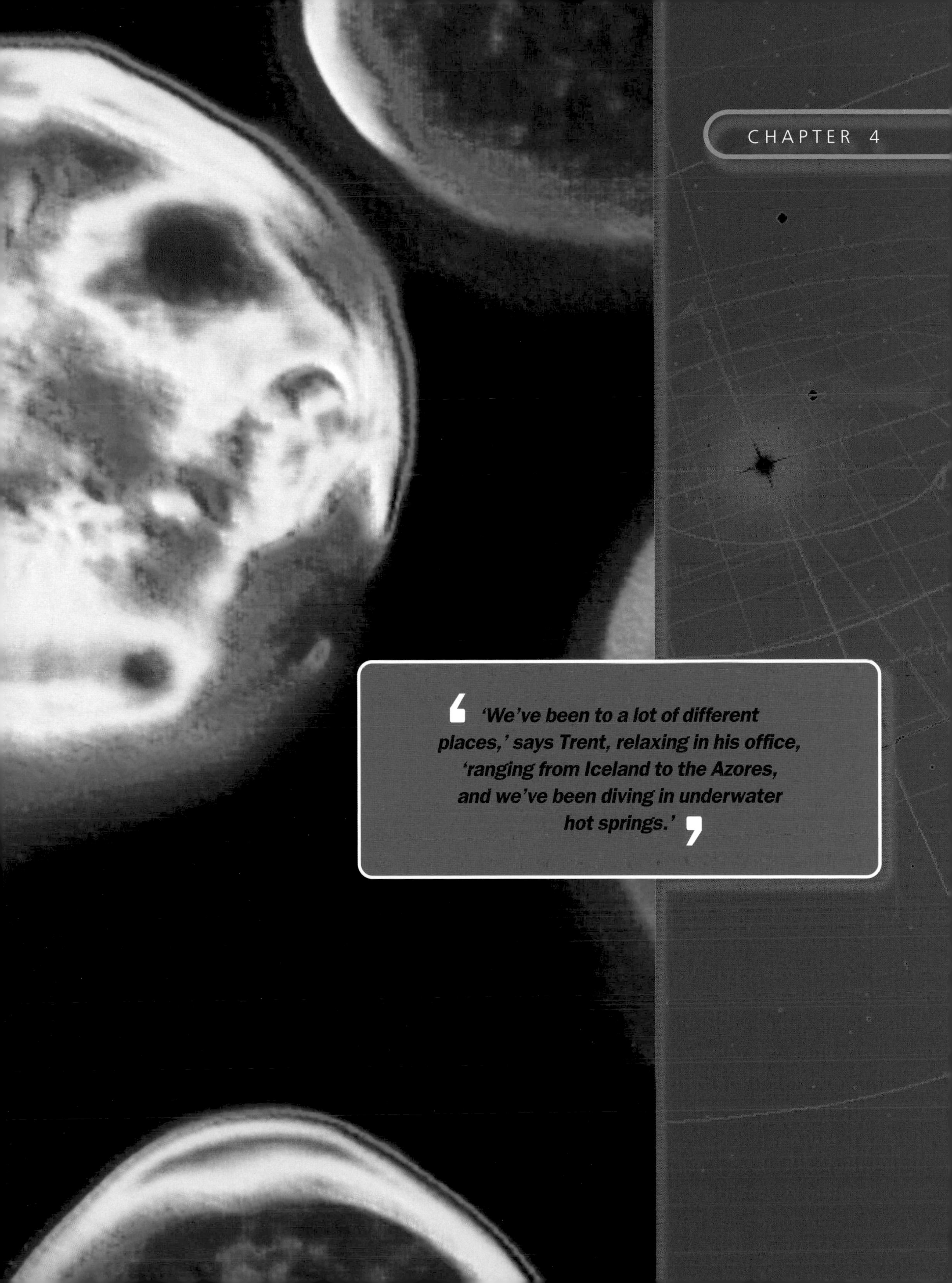
CHAPTER 4
'We've been to a lot of different places,' says Trent, relaxing in his office, 'ranging from Iceland to the Azores, and we've been diving in underwater hot springs.'

Amid the steaming hot springs of Yellowstone Park, Jonathan Trent is literally feeling his way to the strangest life forms on Earth. The tall, bearded biologist is carefully probing the ground ahead with a long pole. Only when he knows there's a firm footing does he gingerly step forward.

In search of the Earth's hottest life-forms, NASA's Jonathan Trent carefully extracts a sample from the steaming poisonous pools of Wyoming's Yellowstone National Park.

'These springs often form very thin crusts,' he explains, 'and if you inadvertently step on these crusts you can end up burning yourself in the scalding water. That's why we probe with these poles, and move around like blind men.'

But burned feet are the least of the hazards here. Fall in completely, and you've worse worries in store. 'The Park Service says that, if you ever do fall in, then you don't want to be pulled out, because the long and excruciating death would be worse than the quick one that would happen if you're just boiled in the springs.'

Trent recalls one person who fell into the springs and survived – for three agonising days. He didn't die of either scalding or the acid environment. It turned out the springs are also filled with arsenic, and he died of kidney failure.

So what drives this enthusiastic young researcher to take his life in his hands? 'We're going to start exploring for life on other planets,' he explains, 'and the very first thing we should be doing is to understand the extent of living things on this planet.'

In the past few years, scientists have discovered life on Earth that can live under the most alien of conditions that our planet can throw at it, from Trent's hot springs to the permanently frozen wastes of Antarctica and the deepest oceans. These are the aptly named 'extremophiles'. Most are no more than microscopic bugs, but they are bugs that can thrive in conditions fatal to advanced life, and where our most sophisticated technology will fail.

His cautious advances have led Trent to an inviting spring – one that's boiling, with superheated steam and spitting acid water. Standing well back, he lowers a video camera into the cauldron. 'These small cameras with built-in lights were originally developed to explore the plumbing in nuclear power plants,' says Trent. With stainless-steel bodies and tough plastic lenses, the cameras are relatively immune to the most corrosive environment on Earth.

'We "improved" them for use in the hot springs by disconnecting the overheating circuits,' Trent chuckles. 'This means we can get the cameras up to 120°C, and we get fairly good images for about an hour.' With long telescopic poles, Trent checks out the temperature through

'The long and excruciating death would be worse than the quick one that would happen if you're just boiled in the springs' the spring. And – most important – he swings out a sample tube and drops it into the heart of the boiling springs.

Life can take in its stride the worst conditions encountered on Earth

Back in the lab, the sample is checked under the microscope. It's swarming with extremophiles: extraordinary bugs that live and survive in the poisonous acid spring, superheated like a pressure cooker to well above the normal boiling point of water.

'We've been to a lot of different places,' says Trent, relaxing in his office, 'ranging from Iceland to the Azores, and we've been diving in underwater hot springs.' He makes a preliminary expedition with his wife Susannah, who videos the location to check what equipment they'll need when they come back in earnest. 'For example, the last trip was to Mount Lassen Volcanic National Park, and we ran into the difficulty that – even though it was June – we had ice fields to cover, so we needed to come back with crampons. And we had some serious climbing to do, so that would restrict what we could carry and really limits the kind of thing we do.'

Human researchers may be limited in these conditions, but microbes are not. Some extremophiles just don't feel the cold. It's not just a matter of containing an in-built antifreeze. Trent has discovered that these organisms make special kinds of proteins that help them to adapt to freezing temperatures. With these proteins, a microbe doesn't just go into suspended animation: the proteins keep the wheels of life turning. Trent enthuses: 'There's just recently been the discovery of a species in Antarctica that lives exclusively at low temperatures – and can still *grow* at temperatures below the freezing point of water.'

Extremophiles can also thrive in ponds as strongly acidic as a car battery; lakes that are equally alkaline; and saltpans where the brine is totally saturated. Life can take in its stride the worst conditions encountered on Earth – and Trent believes that this opens the way to any number of alien environments where life could thrive.

'As we try to explore beyond our planet,' Trent explains, 'we're going to need to

Though deep-frozen and now dry, this old riverbed on Mars – Nanedi Vallis – could be a salubrious habitat for extremophiles that live and breed in Earth's coldest locations in Antarctica.

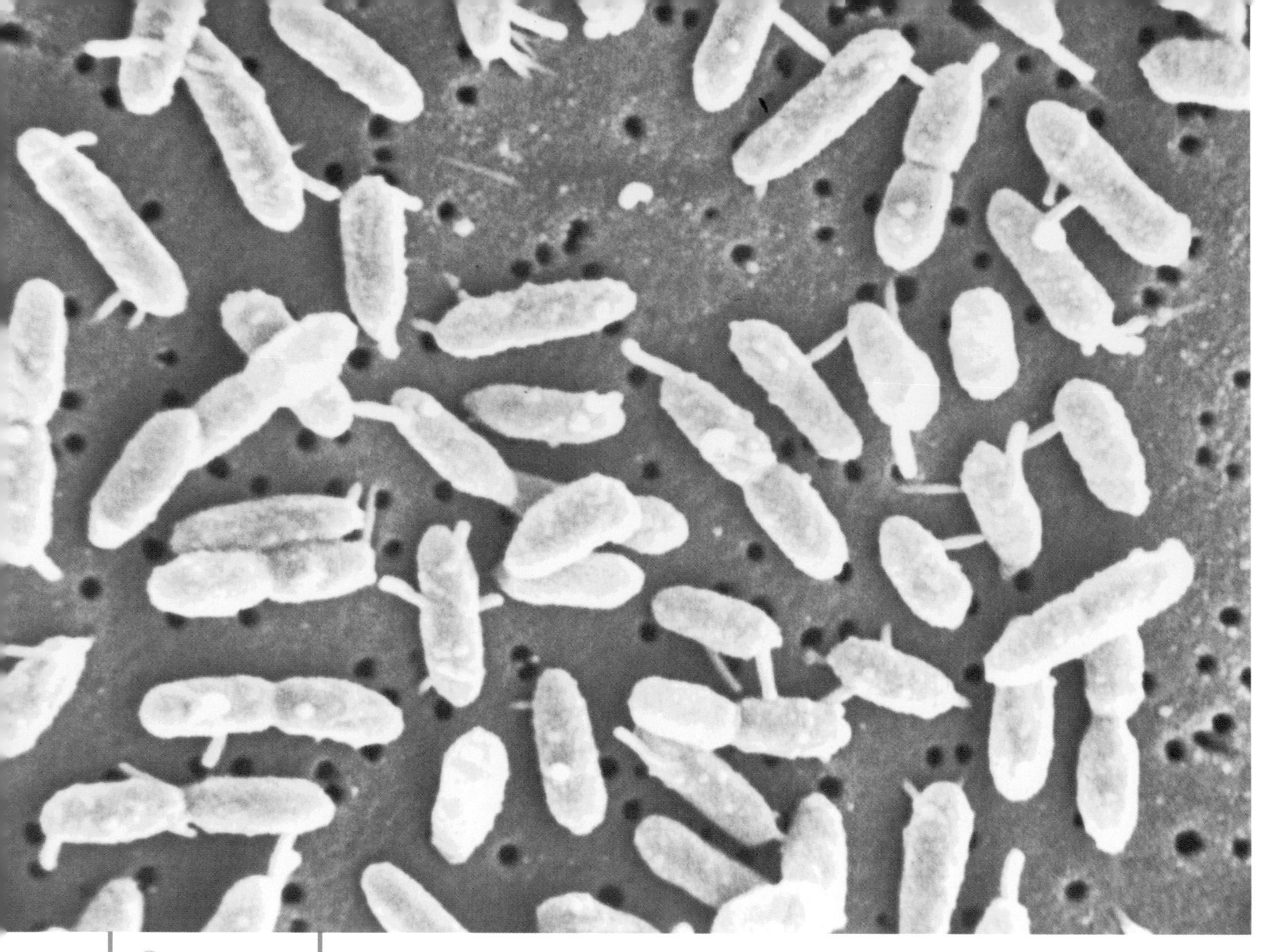

Acid-house bugs! These highly-magnified extremophiles are *Acidiphilium* bacteria, living in the highly acidic waste from mines: some are enjoying it so much they are procreating (splitting in two).

understand not only what might be living there – given the bestiary of organisms we know live here – but also how it is they manage to adapt. We're exploring these extreme environments because it gives us a handle on how things have evolved here, and what modifications of the biochemical machinery that makes up a living system can be made in order to make something function in an extreme environment.'

True to its futuristic aims, the institute itself exists in cyberspace: it's a virtual linking of researchers at a dozen locations, ranging across the US and beyond

That's why Trent is currently working at the headquarters of NASA's rather optimistically named Astrobiology Institute. We say 'optimistically', as the researchers have yet to find the subject of their study: 'biology' in the 'astronomical' realm.

True to its futuristic aims, the institute itself exists in cyberspace: it's a virtual linking of researchers at a dozen locations, ranging across the US and beyond – from the Mexican border to Cape Cod; from the Rocky Mountains to Spain. Appropriately, the hub of this cyber-institution is to be found in California's Silicon Valley.

To visit Jonathan Trent and his colleagues in person, we've arrived at NASA's Ames Research Center. It's the sort of place you can't miss. Silicon Valley is dominated by its vast Hangar 1, almost a quarter of a mile long and 200 feet high. Hangar 1 was built in the 1930s to house the giant navy airship USS *Macon*. A few years later, an aircraft research lab was built next to the airfield. Its striking pre-war buildings now form the heart of the Ames Research Center, where NASA is investigating an astounding array of ideas that will shape the future.

On every previous visit to Ames we've stumbled over something bizarre. Some can't help but be obvious: as well as Hangar 1, Ames boasts the world's biggest wind tunnel. On our first trip, in 1985, we'd been surprised to spot a U2, America's famous 1960s spy plane – now adapted to pick up radiation from deep space. Over the years, we've filmed Ames's flying observatory – a telescope poking through the side of a giant transport plane cruising at 41,000 feet – small robots that float alongside an astronaut as 'personal assistants', and a team that was searching for radio broadcasts from aliens.

This time our quest concerns the inner secrets of Mars: ultimately, we want to know if there is life on the Red Planet. But first we must understand The Meaning of Life. And where better to seek it than the Astrobiology Institute, and who better to ask than the institute's director, Baruch Blumberg?

'The mission statement for astrobiology in general,' explains Blumberg, 'is the study of the origin, evolution, distribution and future of life on Earth and in the Universe.'

The magnitude of this task doesn't appear to faze Blumberg. But then he's overcome the odds many times before in his long career as a research biochemist. In 1976 he was awarded the Nobel Prize for Medicine for discovering the liver-destroying virus hepatitis B – and then developing a vaccine to control the virus. 'The vaccine is now being used very widely,' says Baruch. 'More than a billion doses have been administered over the last ten or fifteen years.'

Baruch's work has already saved millions of lives. And it promises to do more than preventing cases of hepatitis B. Baruch digs out for us a recent report from Taiwan, showing a large decrease in cancer of the liver – one of the most common cancers in the world. He's typically modest about the work. 'That's just what I did prior to going into space,' he jokes.

This time our quest concerns the inner secrets of Mars

And Blumberg's space odyssey has one goal – to check for life out there. So we start at the deep end. 'Just what *is* life?'

'What a fantastic question,' he begins. 'Well, there's a set of characteristics you can use. You need replication and diversity – the fact that you have different kinds of offspring that aren't exactly the same – therefore natural selection can operate on them. You might say there's metabolism: you have to be able to take energy from some place and then convert it into energy within the lifelike phenomenon. And there's more. With the continuation of astrobiology, these definitions will be enlarged and amended.'

Replication of life on Earth is based on DNA – the famous 'double helix' molecule that passes down all the genetic information to your offspring. But Blumberg doesn't see DNA as an essential ingredient of alien life.

'There may very well be other ways of putting molecules together,' he explains, 'that can serve the functions of replication, diversity and so forth.' Warming to his theme, Blumberg explores the philosophy behind the idea. 'The only model we have for life in the Universe is life on Earth – so that's our current model, which means DNA and proteins and so forth. What we're doing is testing this first model, with the expectation that something new and different will come along – something that's very hard to predict right now.'

If alien life is that different, it could require some lateral thinking to track it down. It's time for us to turn to another team of NASA astrobiologists.

Near the city of Pasadena, above the sprawl of Los Angeles, lies the Jet Propulsion Laboratory. Its name reflects its original aim, but not its current preoccupations. JPL is NASA's centre for unmanned spacecraft. The lab has designed and built sophisticated robots that have explored all the planets of our Solar System except remote Pluto.

JPL's experienced space-engineers are now building the next generation of spacecraft to visit Mars. But how are they to recognise life on the Red Planet? They turn to Pam Conrad. As she's the first to admit, 'I have the coolest job in the world – it's to develop strategies and tools that will help us detect life ultimately on some other planet.'

'I didn't always do this kind of thing,' she explains. With a father who had to travel to take part in nuclear tests, Conrad fell behind at school. Unable to keep up in mathematics, she turned to music. 'I've done every possible kind of music except perhaps hip-hop and rap,' she laughs. She also worked as a television producer. 'One day the crew were following me through a dreadful thunderstorm, and we cut through a building in a university. It was the geology building,

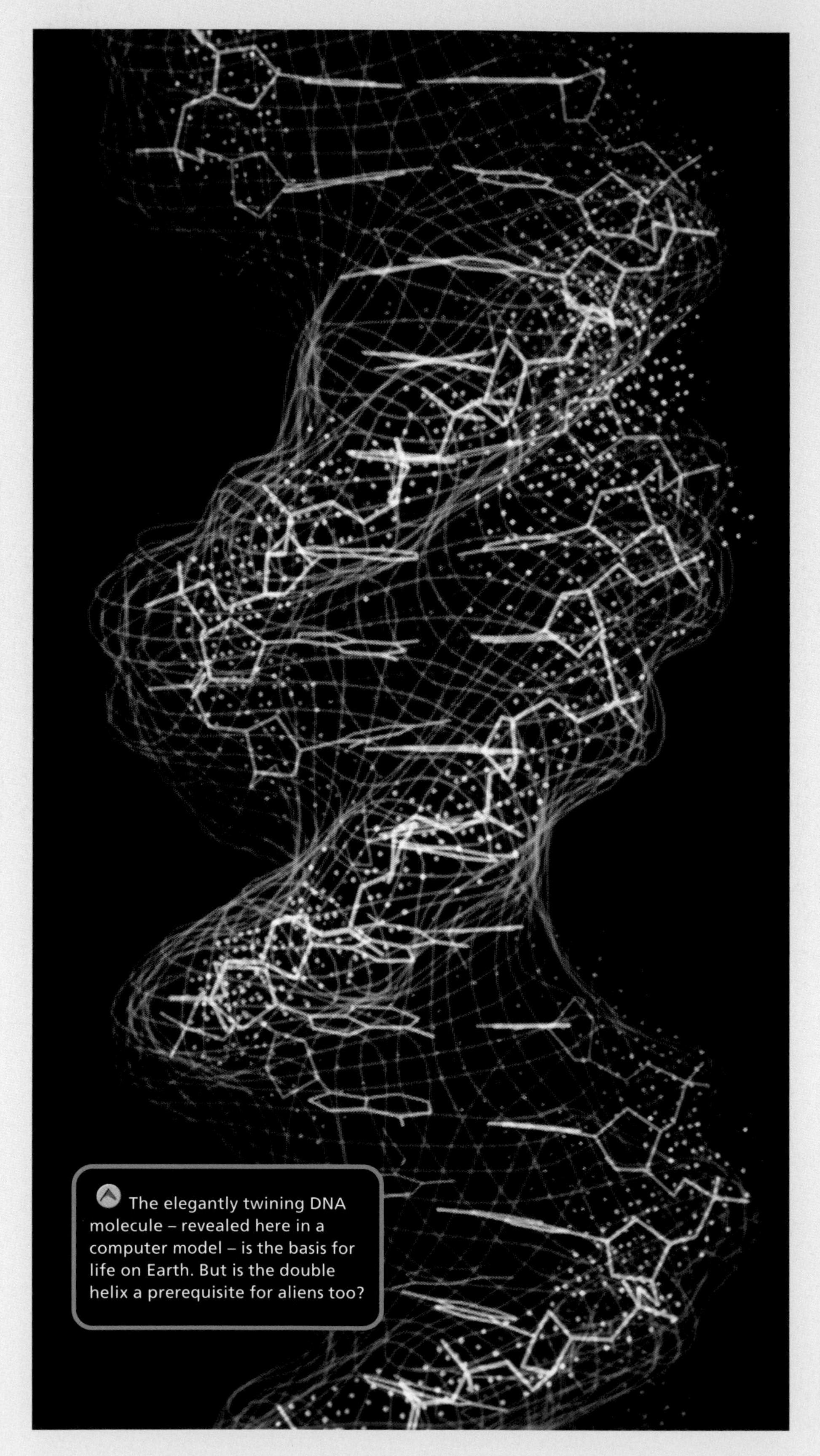

The elegantly twining DNA molecule – revealed here in a computer model – is the basis for life on Earth. But is the double helix a prerequisite for aliens too?

and there were all these beautiful crystals and minerals in a showcase. And I had this flash – this visceral reaction – and I said: I love science.'

For Conrad, the experience reawakened early yearnings for science and space. So she registered for science classes, not caring how rocky the path might be. 'It's like love: you're willing to forgive all kinds of flaws in the beloved if you have that motivation. I said I will do this math, I don't care how hard it is, because I love this. And when you really apply yourself, it's no harder than figuring out what to cook for dinner – it's just another set of skills.'

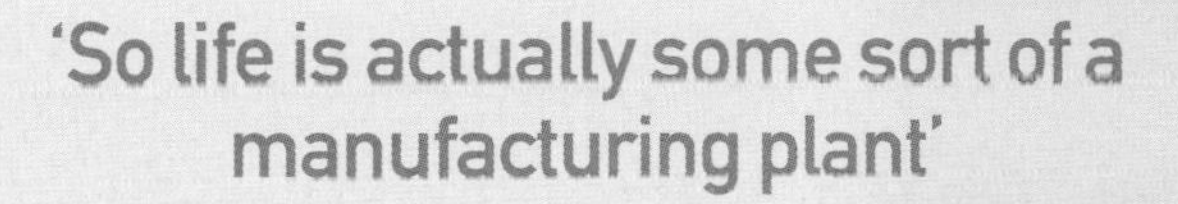

Conrad takes a typically practical attitude to The Meaning of Life. 'One thing that defines life is its ability to take in material and make something, and continue to maintain this process. So life is actually some sort of a manufacturing plant. But in order to be something different from, say, the Ford Motor Company, we have to have some other features of life.'

Another important feature is some kind of structure to support life's processes, and the ability to reproduce that Blumberg has already picked out. 'There's the Darwinian definition,' Conrad expands, 'that not only does life replicate, but it evolves, it changes, it mutates, it responds to changes in its environment and becomes even fitter.

'And we're exploring the chemistry of life,' she concludes. 'What are the constituents that make something look like life, as opposed to this table or a rock or the sky or a swimming pool?'

Our living cells depend on some pretty unique reactions between some unusual chemicals. But why should alien life be made of the same chemicals as us? Perhaps carbon atoms and organic chemistry are a peculiarity of life on Earth? Pam Conrad has some sympathy with our question. 'Well, as any geeky kid, I read many comic books growing up, particularly *Superman*, and I do remember an enemy of Superman who was made out of silicon.'

Some scientists, too, have suggested there could be silicon-based life in the Universe, with a metabolism based on a different range of chemical reactions. But Conrad is sceptical. 'On Earth there is abundant silicon, but when you think about how life does the business of metabolism, it makes more sense to use carbon than silicon. It's energetically less expensive to do this business with carbon than with silicon.

'But we do know this,' she continues. 'We see evidence of organic chemistry everywhere in the

Not our flesh-and-blood... While looking vaguely like humans, these hypothetical aliens are made not from organic matter but from silicon, the semi-metallic element used for computer chips.

Universe, in meteorites, even in the interstellar medium. It's not just on Earth. So that's very helpful. It means that organic chemistry could evolve through pre-biotic chemistry to life in other environments. It lets us know that our assumptions are somewhat good; that organic chemistry is a good pathway to get to life.'

In 1953 American chemist Stanley Miller re-created the early Earth in a glass flask

Our quest for The Meaning of Life is now focusing on organic chemistry – all the comings and goings between molecules that are rich in carbon. To create life on Earth, the only place where we can really study it, our planet must have been blessed early in its life with a profusion of simple organic molecules, the raw material of life.

Astrobiologist Chris Chyba, now in Silicon Valley, has spent years working out where all this gunk came from. And it's a frustrating business.

'Genesis Man' Stanley Miller surveys the contents of a flask that mimics the early Earth. From simple gases and a spark, he's created the raw materials of life.

'You know, forty-five years ago I think everyone would have thought that if there's one part of the puzzle of the origin of life that we really understand, it's the origin of these simple little organics. Now it's the case that we're not even sure where these simple molecules came from.'

In 1953 American chemist Stanley Miller re–created the early Earth in a glass flask. Above a pool of water to represent the early oceans, he introduced an 'atmosphere' of ammonia and methane. Miller sent powerful electric discharges through the gases to mimic lightning, and then went home for the night. When he returned the next morning, the miniature ocean had turned orange. Dissolved in the pool were all kinds of simple organic

The raw material of life is truly universal. Organic molecules pollute interstellar space with vast dark clouds (inset), and nearer home they are scattered across the Solar System by comets like Hale-Bopp, pictured here over California's Mono Lake, a haven for primitive bacteria.

molecules, forged from the atmospheric gases by the power of the lightning discharges.

'It was unquestionably a very important experiment,' Chyba comments. 'But probably the atmosphere he used is not the atmosphere that existed on the early Earth. Now, we think that the Earth's early atmosphere was rich in carbon dioxide. And, in that case, it's very, very much harder to make organic molecules.'

Instead, astrobiologists are beginning to look skywards for our ultimate origins. Biologist Bill Schopf, in Los Angeles, points out our cosmic connection. 'Life is mainly made of four elements. We call them CHON – carbon, hydrogen, oxygen and nitrogen – and these turn out to be four of the five most abundant elements in the whole Universe. As the late Carl Sagan used to say: we are made of star-stuff.'

This star-stuff is on display for us all to see. Look at the Milky Way on a dark night, and you'll spot black patches silhouetted against the glowing starry background. These are dense clouds in space, thick with CHON dust. 'When you study these great vast clouds with radio telescopes,' says Schopf, 'you can identify molecules in them. So you've huge, huge clouds packed with these little molecules.

'This organic matter is carried around in this Universe of ours on comets, like Halley's Comet,' Schopf continues, 'and when one of these comets comes close to the Earth, it could bring water – and also bring organic compounds.'

The proof came in 1986, when a trio of spacecraft flew through the head of Halley's Comet. The two Russian Vega craft and the European Giotto were severely battered by comet dust, but in return they got to analyse comet-stuff at first hand. The dust was mainly organic – chunks of pure CHON.

And we certainly know that these celestial tramps can – and do – hit the Earth. In 1908 a comet or frail asteroid exploded over the Stony Tunguska valley in Siberia, devastating thousands of square miles of forest. And a bigger impact in Mexico, 65 million years ago, delivered the *coup de grâce* to the dinosaurs.

'We've found that you can successfully deliver organic matter to Earth even in big comet impacts,' Chyba continues. 'It's kind of surprising, because the temperatures are so high. But the high temperature pulse lasts for only a short time, and there are regions of the comet that experience only low temperatures. If you put all the numbers together, it looks like cometary delivery – for certain amino acids – could have dominated what was being made in the Earth's atmosphere.'

Amino acids are the building blocks of proteins. And proteins are – along with DNA – the vital molecules for terrestrial life. But comets are not the whole story: our origins may lie in smaller interplanetary fare. Chyba: 'I think it's possible that amino acids delivered in dust may have dominated what you were delivering in big comet impacts.'

On the trail of life-giving dust, we head further south on our long Californian odyssey – to San Diego. We're welcomed at the Scripps Institution of Oceanography by Danny Glavin. 'Would you excuse me for a few minutes first, while I go to the lab?' he enquires politely. 'Otherwise the experiment will have to wait another hour.'

Intrigued, we follow him. Like the school lab of our youth, gleaming glassware is stacked everywhere, but the

'If you put all the numbers together, it looks like cometary delivery... could have dominated what was being made in the Earth's atmosphere'

It could be three billion years ago: stromatolite colonies dappled by sunlight shining through the waves of Western Australia. These bacterial mounds are living relatives of the Earth's earliest life-forms, found fossilised just a short distance inland.

experiment in question is encompassed in a stainless-steel cylinder hooked up to electronic equipment. Glavin injects a sample, and leaves the experiment to run itself.

'What we have in there is some interplanetary dust,' he explains, 'collected from the Antarctic ice sheet. We melt the ice and filter out these tiny grains – they're real small dust particles, hard to see with the naked eye.' The equipment running in the background is dissolving the grains and analysing them for amino acids.

'The basic idea,' Glavin continues, 'is that the early Earth was seeded with the organics necessary for the origin of life. And what you have to remember is that these tiny dust grains provide the bulk of the mass accreted to the Earth from space each year. I mean hundreds or thousands of tons of the stuff are bombarding the Earth.'

Raining down through the atmosphere as microscopic shooting stars, the dust particles are heated until they glow red-hot. Scientists had thought that the miniature inferno would destroy any organics being carried aboard.

But Glavin now knows different. 'I've actually tried to model this process in the lab by heating some of these meteorite particles to 1,000°C. And what I'm finding is some of these amino acids will vaporise from the surface of the grain into a gas, and then re-condense in the colder region behind the meteorite. So you get amino acids essentially raining down to Earth.

'I should make it clear,' Glavin concludes, 'that I'm not saying that extraterrestrial delivery was the only way that the Earth was getting seeded with the building blocks of life. The Miller process is another great way to make amino acids. But the bottom line is: any way you cut it, you're getting organics to the early Earth.'

Schopf concurs: 'There are many sources of the four essential elements – the CHON molecules – on the early Earth. You can make it in the primitive atmosphere; it could come from comets; and there are meteorites that contain 4, 5 or even 6 per cent of these molecules. So organic matter is really common throughout the Universe; the Earth certainly had it when life started up.'

To us, it still seems one giant leap to get from organic gunk to living cells. But that doesn't faze Bruce Jakosky, a geologist-turned-astrobiologist. 'We see that life originated on Earth very quickly,' Jakosky explains, 'and that to me says that the origin of life is a very straightforward natural consequence of simple chemical reactions in a planetary environment. If we try to simplify it down, all we need are three things: the availability of the right elements, liquid water and a source of energy.'

Bill Schopf doesn't seem too concerned either. He is famed for discovering the oldest cells on Earth – microscopic fossils from the Australian outback, which lived 3,465 million years ago. 'It's a desolate place,' he recounts, 'with a few kangaroos, a few emus and a few stray cows and nothing else, except scrub-brush and desert. You collect rocks, bring them back to the laboratory and then – if you're really lucky and work hard – you're able to find these fossils of microscopic organisms.'

'The total mass of life below the Earth's surface is considerably greater than the amount of us creatures walking around up on the surface'

If anyone has a vested interest in the origin of these first cells, it must be Schopf. And he envisages a straightforward process. 'So we've made the little molecules, like amino acids, and they get linked together like beads on a string to make a protein. And we've also made little snippets of the information-containing molecules, like DNA, and they link together too. And the organic matter got together as little droplets within the surrounding water, just like oil balls up into drops in a bucket of water – and that's where cells came from.'

Optimistically, Schopf sums up: 'So we understand the broad outlines of the picture; what we don't understand are a whole lot of the details, and that's why science is busy working on it.' And for us, the devil is still in the detail. Not just how the mechanisms of life got going, but where?

Schopf envisages a warm, shallow sea, with the first cells living like pond scum. In time, they came to build up into thick colonies, like rocky cushions, called stromatolites. Amazingly, Western Australia boasts not only fossilised stromatolites from the dawn of life on Earth, but also stromatolites still living, in the warm waters of Shark Bay.

Others aren't so sure. Bill Hartmann is a geologist studying the hottest places on Earth – thermal springs, geysers and volcanoes. 'Rather than what people thought twenty years ago, which were these balmy tropical pools, it now looks as though the earliest cells formed in hot springs.'

He bases his assessment on the DNA found in living cells. Biologists can construct a family tree from the DNA, working out how close we are to the chimps, for example, and also how long ago we diverged from our ape-like brethren. Pushing the technique back in time, they can hope to tell which were the earliest kinds of cell on Earth.

'When you use DNA evidence and go back in time,' Hartmann explains, 'to find

A hydrothermal vent, on the floor of the Pacific Ocean, belches superheated water and minerals into the cold ocean depths. Exotic life-forms – unique on Earth – thrive round these 'black smokers'.

our common ancestor on the early Earth, it turns out to be extremophiles – these bacteria and microbes living in very high temperatures. They still live on: you go to Yellowstone Park and you see them, in colours that show a colony tuned to a particular temperature range.'

But Baruch Blumberg is sceptical of any habitat open to the sky. 'There was a great deal of volcanism and many impacts from meteorites and larger bodies. Life could not have survived that kind of onslaught unless it was in some location deep under the Earth or the sea.'

Subterranean life sounds like something from a horror movie – yet it's now been found to exist. In the mid-1980s, geologists in the US drilled over a mile deep into the basalt rocks near the Columbia River. To their surprise, the water from the depths of the hole was swarming with bacteria – and bacteria very different from those that live at the Earth's surface. These, too, are extremophiles. They live in tiny cracks in the solid rocks deep inside our planet; they feed on chemicals dissolved from the rock; and they live and breed at an extraordinarily slow rate.

Bill Hartmann puts these microbes into context: 'It's been calculated that the total mass of life below the Earth's surface is considerably greater than the amount of us creatures walking around up on the surface – and, you know, we always thought we were the main show in town!'

Cracks in the Earth's rocks could have provided the ideal cradle for life – microscopic test tubes for assembling the amino acids and DNA fragments into whole cells, safely protected from anything the Universe could throw at them.

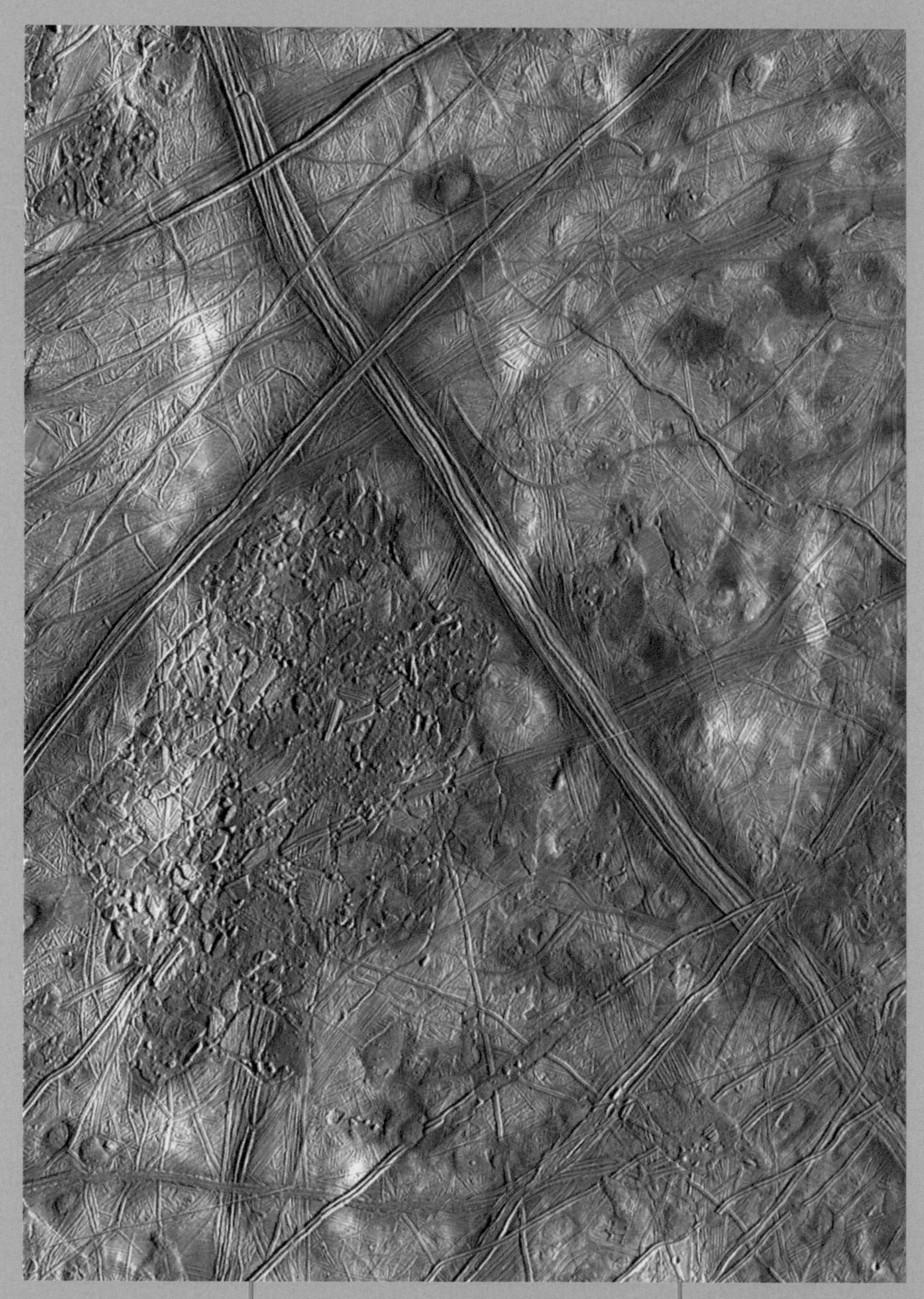

The cracked surface of Jupiter's moon Europa is a thick layer of ice, concealing a deep ocean beneath. Here we may find life in an environment even stranger than anything faced by life on Mars.

On the other hand, extremophile expert Jonathan Trent believes the oceans would have been an even safer haven. 'Clearly, life originated as a microbial-like beast, but I'm really lost to come up with a good description of where it would have been. It must have been protected from the bombardment and from the radiation in the Earth's early history.' One scenario is that life evolved in a cold context. 'Most of the inhabitants of this planet are microbes living in the deep sea – and the deep sea is a cold, dark environment.'

Many scientists are taking the prospect of life on Europa very seriously. And if life could start on such an alien world, then our neighbour planet Mars suddenly seems an absolute Eden

Frank Drake, the pioneer of radio searches for distant civilisations, is intrigued instead by the *hot* places in the Earth's oceans. At deep-sea vents, volcanic heat is sending streams of boiling water upwards from the ocean floor. 'It now looks as though life could have – and perhaps did – originate at the deep-sea vents, and those will surely be more common than the old idea of warm little ponds.'

The prospect enthuses Drake, ever an optimist when it comes to life and intelligence throughout the Universe. 'So that encourages the idea of life developing in many, many places – places we, a few years ago, would have thought impossible.'

One such place is Jupiter's moon, Europa. Beneath a crust of solid ice, Europa almost certainly harbours a deep ocean – containing something like twice the amount of water we have on Earth. And most scientists believe that the ocean floor is riddled with deep-sea vents spewing boiling water. The evidence is twofold. Without a fire beneath, Europa's ocean would long ago have frozen solid. And Europa's twin – but dry – moon, Io, is a hotbed of erupting volcanoes.

Perhaps life could exist at Europa's deep-sea vents? Pam Conrad is excited. 'Europa's an interesting place. There's this evidence for hydrothermal vents, so perhaps there is some life exploiting them – like the organisms in Earth's deep-sea vents, which breathe metal.'

But Chris Chyba is more cautious. 'Most organisms at hydrothermal vents on Earth depend on oxygen, which has come down from the surface.' Europa has no atmosphere, and no obvious source of oxygen. 'So even if there are hydrothermal vents,' Chyba continues, 'will anyone be able to make a living there? You know, it's quite possible that we'll get there and discover an ocean's worth of water that is utterly sterile. That would be a startling result from an Earth perspective, where we think of the ocean as the cradle of life.'

Despite Chyba's scepticism, many scientists are taking the prospect of life on Europa very seriously. And if life could start on such an alien world, then our neighbour planet Mars suddenly seems an absolute Eden.

Chyba warms to this idea. 'Life may well have originated on Mars, especially since there's this evidence for liquid water early in Martian history.' And it strikes a chord with Bill Hartmann too: 'If these things are forming in hot springs, and we have volcanism on Mars throughout its history – then why not imagine that the same processes produced microbial life there?'

In fact, whatever the cradle of life on Earth – hot springs, warm pools, deep cracks, cold oceans or boiling ocean vents – almost certainly you'd find the same on Mars in its earliest and most active years. And Mars

No, it's not an alien bio-spaceship out among the stars, but a plankton found floating in the Earth's oceans. Life on Europa could be far more exotic still.

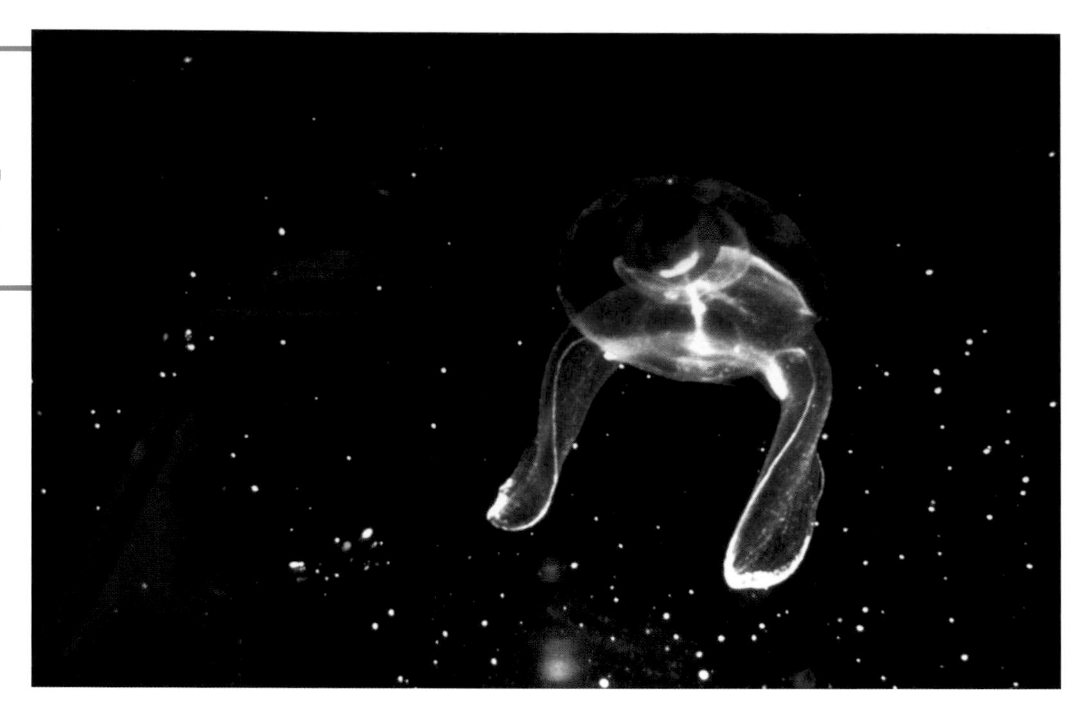

would have received much the same cornucopia of organic molecules as our planet.

'At the beginning of its history,' says Nathalie Cabrol of NASA's Astrobiology Institute, 'Mars was kind of like the Earth. Whether the bricks of life were brought to the Earth with the pieces that formed the planets, or some other way, there is a great chance that the same processes happened on Mars.'

'We now know that Mars is just like the Earth,' adds Danny Glavin, 'getting organics from meteorites. So even if Mars wasn't able to produce the organics there by itself, it was certainly getting an extraterrestrial input, which could have helped kick this evolution of life there. I think the probability is really good that life at least tried to get started on Mars.'

Even Bill Schopf will concede the possibility – and he's been a major sceptic over claims that the Martian meteorite ALH 84001 contains fossils of life from the Red Planet. 'It's plausible that life could have begun on Mars, when Mars was more like the Earth. That's certainly possible.'

The big question now is: has life survived on Mars to the present day?

There's certainly no evidence for anything bigger than microbes. SETI pioneer Frank Drake believes that's just an unlucky accident of the planet's birth. 'I think the Solar System got cheated. Mars should have been a bigger planet, and had it been a little bigger it would have been very suitable for life and we could well have had two planets in the Solar System with very highly evolved life on them. Boy, that would really cause tremendous interest in extraterrestrial life!'

But Mars's fate was to be a frozen world, under a thin atmosphere with no protective ozone layer

'There are ten times more bacterial cells on our bodies than there are human cells; our skins are covered with microbes and our intestines are filled with them. So we are basically walking microbial colonies'

to shelter its icy deserts from the Sun's deadly ultraviolet rays. There was no chance of life evolving into anything as complex as a slug, let alone an alien intelligence.

'Life is tough to extinguish once it gets started'

Extremophile expert Jonathan Trent is keen to get such thoughts into perspective. 'When we call these environments "extreme", it's only extreme relative to us – and we're pretty puny when it comes right down to it. We can live in only very restricted environments, and our body temperatures are very limited.

'For the first 2 billion years of life on Earth,' he continues, 'microbes were the whole story, and they continue to dominate the planet. In fact, even human life is completely dependent on microbes. There are ten times more bacterial cells on our bodies than there are human cells; our skins are covered with microbes and our intestines are filled with them. So we are basically walking microbial colonies.'

Put this way, a Mars covered with bugs seems a more attractive proposition. And astrobiologists are excited that Mars may provide habitats as enticing in their own way as our intestines.

'To me it's very likely that there is existing bacterial life on Mars,' opines Drake, 'and it's subterranean. What we've learned from the Earth is that life can thrive deep in the interior of a planet.' Eking out a slow existence in cracks deep below the surface, Martian bugs could live on a diet of chemicals leached out of the rocks.

Trent believes it would fit in with the way in which extremophiles live on Earth. 'It's not inconceivable that Mars could have a very slowly turning over ecosystem, deep below the surface where there's liquid water due to the heat from Mars's interior.'

The only dissenting voice we hear is that of British chemist Jim Lovelock, pioneer of the Gaia idea – that all life on Earth, along with the atmosphere and the oceans, forms one planet-wide living being. He thinks it's unlikely that even extremophiles can exist on Mars today.

'I always feel that extremophiles are very like me,' he says. 'I exist as an eccentric because I live in a rich and comfortable society and they support odds and ends like me. It's the same with extremophiles – they wouldn't be here but for burgeoning life on Earth.'

The extremophiles in hot springs and salt lakes, for instance, depend on gases in the Earth's atmosphere – particularly oxygen – which are recycled by plants and animals. 'They couldn't manage to live on their own,' Lovelock asserts, 'and Mars being what it is – a violently oxidising, destructive surface and atmosphere – it's a pretty poor place to go looking for life.'

But his is a minority view. More typical is an assessment by Everett Gibson, who reported fossils of bacteria in a Martian meteorite. 'We know one thing from studying the Earth: organisms seek places where they're happy, including subsurface habitats. And we're finding that life is tough to extinguish once it gets started. So there may be biological activity beneath the surface of Mars that is still there today.

'They'd be in areas where there's an abundance of water and the ground is fractured by meteorite impacts,' Gibson continues, 'so the water could be moving around.'

This is home territory to Charlie Cockell, a British biologist who's been investigating a crater smashed out by a giant meteorite in the frozen Canadian Arctic. 'Here we find cyanobacteria eking out a living in the cracks in the rocks: they more readily colonise the shocked rocks than the unshocked rocks.'

His colleague Pascal Lee adds: 'Beyond the polar bears and the muskrats and the caribou, you don't see much life around you. But if you shatter

'My feeling...is all you need is water and some temperatures above freezing – and then you can put life on the surface of Mars'

a rock and prise it open, you will find bacteria teeming in the nooks and crannies. You find them in rocks that would not normally contain these kinds of microbes, because they are normally too dense, too compact. Yet after the impact has heavily fractured and shocked the rocks, and vaporised some of the minerals, then the rock becomes a sponge – and that sponge has become the home to all these microbial colonies.'

To their surprise, Cockell and Lee have also discovered bacteria on the outside of the rocks – with a natural sunscreen, a ball of gel that protects them from the ultraviolet radiation. Cockell has continued this research in Antarctica, under the ozone hole where the Sun's ultraviolet reaches maximum intensity on Earth.

His research has driven Cockell to the controversial view that life may exist not only sheltered within Martian rocks but right out on the Martian surface – even though Mars has no ozone at all, and the surface is blasted with ultraviolet rays a thousand times more deadly even than Antarctica.

'In our field trips,' he explains, 'we see microbial mats where the organics in the upper layer screen out the ultraviolet radiation and protect the microbes below. Now, on Mars you could have living microbes protected under a dead layer of microbes or a very thin layer of dust.'

At the British Antarctic Survey in Cambridge, Cockell is growing bacterial mats under an intense sun lamp to mimic the radiation damage they'd face on Mars. 'My feeling,' he says, 'is all you need is water and some temperatures above freezing – and then you can put life on the surface of Mars.'

One of his hardy bacteria is *Deinococcus radiodurans* – a bug that Jonathan Trent also has under the microscope. Trent explains: '*Deinococcus radiodurans* is the organism that has the greatest radiation resistance of any organism discovered on Earth.'

This microbe first turned up in the 1950s, during experiments in sterilising food by irradiation. Even after they'd been blasted by radiation, some cans of meat continued to go off. The culprit turned out to be this hardy bug. The US Department of Energy is now trying to harness *Deinococcus radiodurans* to clean up nuclear waste-dumps.

Expose a human cell to radiation, and its DNA is blitzed to pieces. Unable to reproduce or even to function properly, the cell dies. 'But *Deinococcus* can repair its DNA,' says Trent. 'Cut its DNA up into tiny little pieces, and *Deinococcus* completely repairs its DNA in a matter of hours, so it can cope with radiation levels thousands of times higher than most organisms can cope with.'

Trent envisages stowaway *Deinococcus radiodurans* as a permanent resident of the International Space Station, and a survivor on even the longest interplanetary journeys. 'I should add,' Trent emphasises, 'that *Deinococcus* has evolved this capacity to repair its DNA probably not to do space travel but because it goes through a lot of cycles of dehydration and rehydration. During the process of dehydration, it runs the risk of DNA damage. So it's basically just over-engineered its DNA repair system – and that's what allows it to cope with radiation.'

A bug like *Deinococcus radiodurans* may even be able to survive an attack from the harsh chemicals that lace Mars's exposed soil. And – whether it's a relative of *Deinococcus radiodurans* or not – the thought of living microbes on Mars today is sending waves of excitement through the world of astrobiology.

> **'So the greatest scientific discovery of the millennium...would be the discovery of *no* life on Mars!'**

At the Scripps Institution in San Diego, Danny Glavin enthuses: 'Our goal is to figure out how life evolved on Earth, and if it ever evolved on other planets like Mars. Mars is the hot topic right now.'

Bruce Jakosky in Boulder, Colorado, says: 'The types of environments where we've found life under extreme conditions on Earth are the same types of environments we think occur on Mars – places like hydrothermal systems, places where ice is present and temperatures very seldom rise above zero. So there's more reason than ever to think that life could exist there.'

'I really think that if life appeared on Mars,' observes Nathalie Cabrol, 'then it's still there. Life has an incredible potential of survival. In incredibly hostile and extreme environments, like the polar regions, it's completely barren, with winds and blizzards. Then all of a sudden you find a tiny depression and less wind, it's a bit warmer, and you will see a flower.' She adds with a smile: 'So I think there's a great chance that there are some tiny Martian bugs laughing at us trying to find them!'

Charlie Cockell is an optimist too – but he has a rather different take on the importance of discovering life on Mars. Cockell reminds us that rocks from Mars have been blasted off the planet throughout its history, and some – like the famous ALH 84001 – have landed on Earth. Each is a spaceship for transporting bugs, such as the radiation-hardy *Deinococcus radiodurans,* from one planet to the other.

'I personally think,' he says, 'that there probably was life on Mars, because the early Earth was covered in abundant microbial communities. At the same time Mars had water and roughly the same atmospheric composition as the Earth, and – as I said – rocks were being transferred between Earth and Mars. My view is that if we find life on Mars it'll actually be quite boring – because it'll be what we expect, based on existing evidence.'

He sees the media as promoting a misguided agenda: 'They see it as "the search for life on Mars", and if we don't find anything on Mars we're all going to go back to our institutions and have our funding cut.

'I think it's really important to understand we're not going to Mars to look for life; we're going there to ask the scientific question: was there life on Mars or not? And it doesn't matter what the answer is. Either way, it's going to be an incredible scientific discovery.'

Suppose the answer is in the negative. That would tell us that we have two planets exchanging rocks and microbes with one another – yet one is seething with life and the other one has remained completely sterile. How in the Universe could that possibly happen?

'So the greatest scientific discovery of the millennium,' Cockell concludes, 'would be the discovery of *no* life on Mars!'

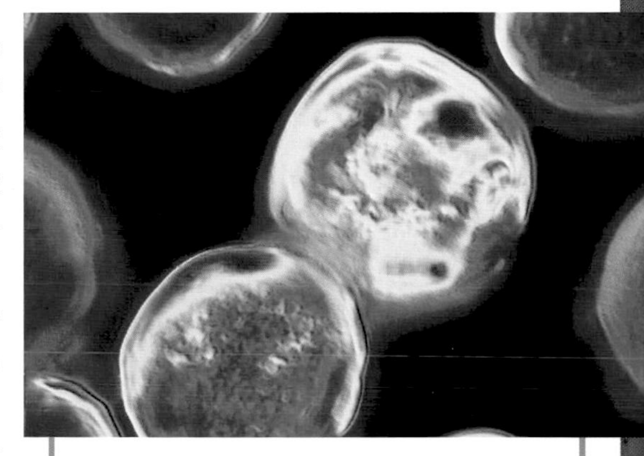

Some bugs like it hot: these *Archaeoglobus* bacteria thrive in the scalding conditions of hydrothermal vents on the ocean floor. Their family tree suggests these bacteria are close relatives of the oldest life on Earth.

THE VIKING INVASION

CHAPTER 5

'I'm an explorer. And so the chance to see something for the first time – a new place – that's where the geology came in.'

It's an unlikely place to be searching for evidence of life on Mars, we reflect, as we pull up at a light industrial estate in Beltsville, Maryland. Anonymous Portakabin-style offices dot the flat landscape, enlivened by vigorous metal sculptures made from industrial spare parts. We walk into Biospherics Incorporated, where we're greeted by a friendly receptionist.

A man in his seventies ushers us into the conference room. 'I'm Gil Levin, founder, president and CEO of Biospherics Incorporated. I was also a member of the Viking spacecraft team. I developed a method to detect micro-organisms very quickly, as a result of a need I saw when I was a public health engineer. I thought this method might work on Mars as well, and so I took it to NASA. On Viking, I was an investigator on the Labelled Release experiment. That's the one that got a positive indication for life on Mars – and has kept me in trouble ever since.'

He can say that again. Viking went to Mars in 1976, and a quarter of a century later some outspoken members of the planetary science community take Levin's claim to have discovered life on Mars with a very large pinch of salt. 'Gil's a sanitary engineer, not a biologist,' says one. Veteran planetary researcher Bruce Murray was director of the Jet Propulsion Laboratory (JPL) when the Viking missions were up and running. We ask for his view on Levin's experiment. 'Flawed deeply. And that's the view of all the people concerned except one guy who's not rational on the subject – Gil Levin.'

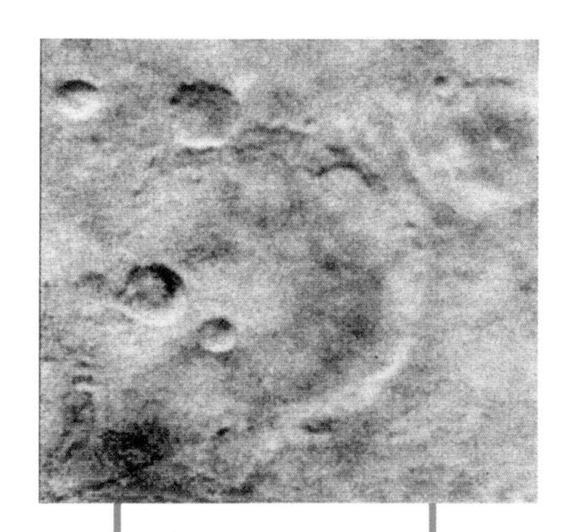

In July 1965, the NASA probe Mariner 4 captured this view of a desolate and cratered Mars from a distance of just over 6,000 miles.

Bruce Murray is one of the pioneers of planetary exploration. Sixty-nine years old, he has a gung-ho attitude to life. 'I remember the Second World War very well. I wasn't old enough to participate, but I was old enough to understand who Hitler was, and what was going on. And the result was that I had a very different view on life – and my colleagues and cohorts did also – that science is great, and that I'm really fortunate to have a chance to do it. Life is full of opportunities, challenges and disappointments – but if you saw something really great to do, you did it.'

Murray started off as a geologist and came to Mars later in life. 'I'm very deeply interested in the Earth and always have been, and so Mars was the consequence of my recognising in 1960 in the United States that something really unusual was happening. And so I changed careers from geology and geophysics – I'd also worked for an oil company and been in the Air Force – then I came to Caltech cold turkey as a post-doc to work on, quote, Space.'

At Caltech (the California Institute of Technology in Pasadena), Murray was let loose on the Institute's observatory on Palomar Mountain – which then housed the biggest telescope in the world. His growing skill in astronomy led to his being picked as a junior member of America's first Mars probe, Mariner 4. The spacecraft flew past Mars in the summer of 1965 and returned pictures of a desolate, cratered world much more like the Moon than the Earth. For a generation of scientists raised on the notion that Mars was Earth-like – and might even harbour primitive vegetation – the images were a bitter blow.

But not to Bruce Murray. 'I'm an explorer. And so the chance to see something for the first time – a new place – that's where the geology came in.'

Mariners 6 and 7 followed Mariner 4, and gave scientists no reason for believing that Mars

> The spacecraft flew past Mars in the summer of 1965 and returned pictures of a desolate, cratered world much more like the Moon than the Earth

was anything more than a larger version of the Moon. Then in 1971 came Mariner 9. The first spacecraft to go into orbit around the Red Planet, Mariner 9 was scheduled to make a one-year detailed photographic survey of Mars. But things didn't quite turn out that way. When the probe arrived, Mars was enveloped in a planet-wide dust storm, which had completely obliterated the surface.

As scientists waited patiently for the pall of dust to clear, Bruce Murray recalls his first impressions. 'Even when the dust storm was there, there were these four dark spots. South spot, middle spot, north spot, and stuff like that.' As the dust storm began to settle, the spots became clearer. 'I thought they were

NASA's Mariner 7 probe flew just 2,000 miles over Mars's south pole in 1969. It revealed yet more bleak and unhospitable terrain, including ice-filled craters.

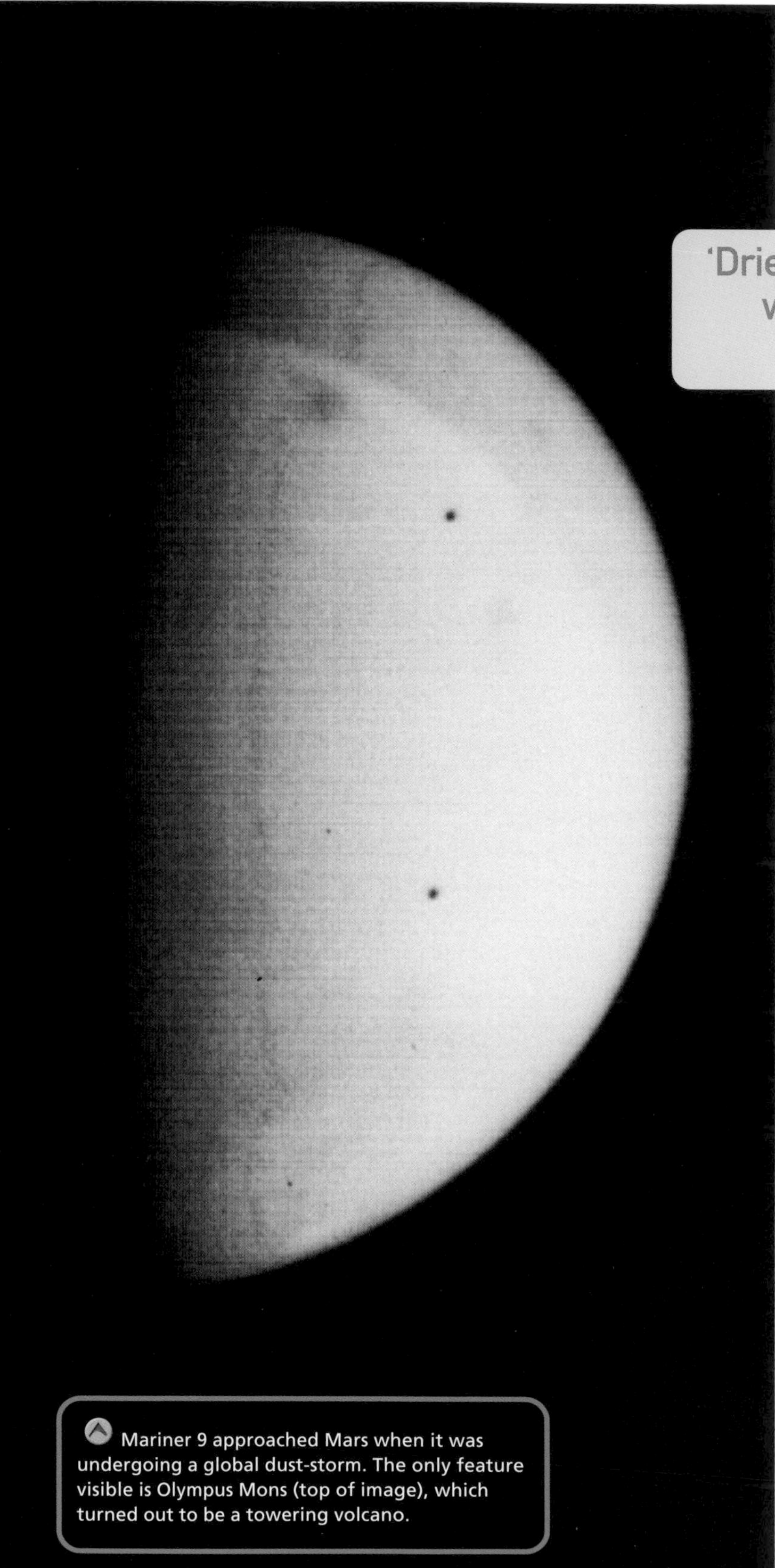

Mariner 9 approached Mars when it was undergoing a global dust-storm. The only feature visible is Olympus Mons (top of image), which turned out to be a towering volcano.

craters initially. It wasn't a good idea, but that's what I thought. In fact, I was involved in a dispute with the US Geological Survey geologists who recognised that these were giant calderas.'

'Dried-up river channels...meant water, that meant life, and therefore all the bugs'

The Geological Survey was right. The 'spots' are huge volcanic cones whose tops have caved in – like the volcanoes on Hawaii. Murray acknowledges his mistake. 'They stood up above the dust. They're really high – the largest is twenty-nine kilometres above the surface. The importance of this was that it meant volcanism on a scale much grander than on Earth. It was hard to swallow, and they were right, of course, and I recognised that after they brought it to my attention. But I didn't grab it.'

Murray ranks Mariner 9 as a ground-breaking mission. 'Mariner 9 told us about the planet and its extraordinarily diverse history. It was the big discovery mission for me and for many other scientists.'

Even as Mariner 9 was circling Mars, another, far more ambitious mission was in its planning stages – Viking. Its motivation was largely political. Murray recalls: 'NASA could already see the end of the Apollo programme in terms of the production of Saturn Vs [the huge launch vehicle that sent astronauts to the Moon]. In 1966 they wanted to do a Saturn V to Mars. A Saturn V could have launched 50,000 pounds, a factor of a hundred increase in mass as compared to Mariner 4. It was crazy, and I thought that it was just absurd, but they were trying to sell Saturn V.' In the event, funding for the proposed mission fell

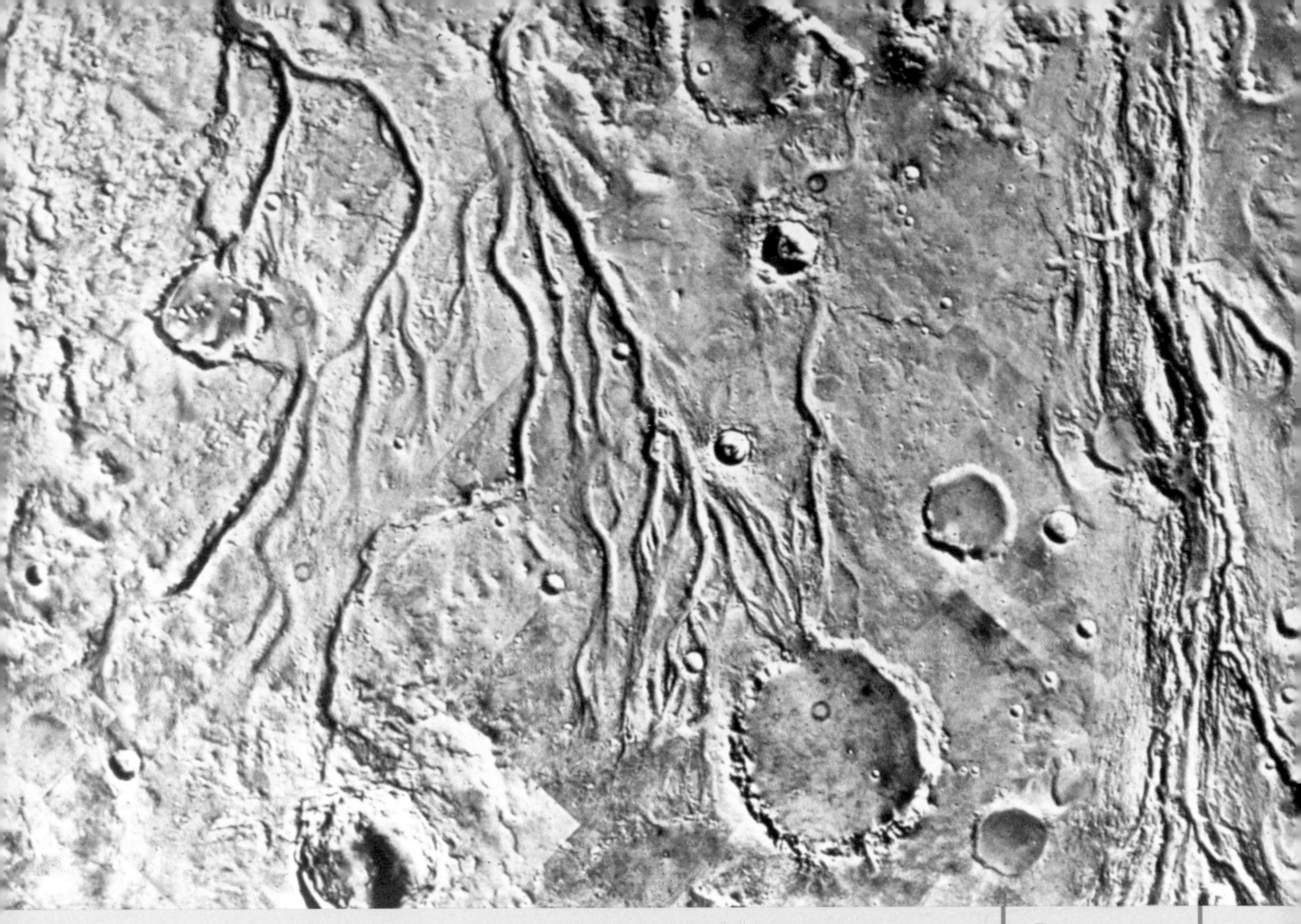

When the dust cleared, Mariner 9 revealed a very different Mars – a world of dried-up riverbeds and braided channels.

apart in the late 1960s. But there was still felt to be a need for a large and ostentatious mission to follow the Apollo programme. 'In mid-1970s money, Viking cost about $2 billion, including launch,' remembers Bruce Murray. 'It was a huge amount of money, larger than any other unmanned programme by a lot. That same amount of money could have been broken up into dozens of smaller programmes. Why doesn't that happen? Because smaller programmes do not yield the same political benefit in Congress as one great big one.'

Viking also provoked a power struggle between the biological community and the more mainstream fraternity of planetary scientists. 'The biologists had the bit between their teeth about life being there,' complains Murray. 'The engineers building Viking said: "OK, if that's what you want, this is what it takes to do it. We'll have to heat-sterilise the whole thing, put it in a bio-shield, and so on and so on." That's why it cost $2 billion.'

Jim Lovelock, a British independent scientist who was invited by NASA to work on Viking, recalls the skirmish. 'There was a sort of minor battle within NASA Headquarters at the time. I wanted to send an atmospheric analysis to look for life, and a surface experiment of a similar kind. But we also wanted to include on the mission as complete a look as we could at the geophysics and geochemistry of Mars.

'But the biologists ran the other faction. They said: "No, no; by far the most important thing is to discover life on Mars. We're practically certain that it's there, and we must go and find it." And they won.'

Geologist Mike Carr, who headed up the Viking Orbiter imaging team, clearly remembers the arguments put forward by the biologists. 'It was those telescopic observations of what appeared to be water clouds – we thought there was water there in the atmosphere. There were seasonal changes that could have been interpreted as biological. And of all the planets in the Solar System, Mars is most like the Earth. This caught the imagination of some very, very influential people: Josh Lederberg at Stanford – a Nobel Prize-winner in genetics – and Carl Sagan.'

Sagan's influence on the Viking project was enormous. 'He was a big player,' admits Jim Lovelock. 'Carl was a nice man – I'll call him a friend. We got on very well, and he was very helpful to me. But he disagreed with nearly everything I said, so he was on the biologists' side.' When Mariner 9 not only discovered huge volcanoes but evidence for dried-up river channels on the planet, Sagan pushed even harder. 'Carl took off into orbit,' recalls Bruce Murray. 'Because that meant water, that meant life, and therefore all the bugs.'

Carl Sagan died in 1996 of bone-marrow disease, aged only sixty-two. He had a glittering academic career. Starting off in the 1950s by researching the origins of life, he made major breakthroughs early in his career by showing that Venus was an inferno, while Mars was a cold desert. Later, he became involved with the SETI community, the group of astronomers searching for extraterrestrial intelligence by monitoring possible artificial radio signals from space.

[Carl Sagan] made major breakthroughs early in his career by showing that Venus was an inferno, while Mars was a cold desert

But it is through *Cosmos* that Sagan will be remembered. The hit TV series and book, which came out in 1980, immortalised the phrase 'billions and billions' and turned Carl Sagan into an international celebrity.

We only got to meet Sagan after *Cosmos*, which was a pity. He struck us as being arrogant and egocentric, but we were delighted to discover that the Mars and bioastronomy communities knew a very different man. Bruce Murray recalls him fondly: 'He was a couple of years younger than me. He did his thesis on the atmosphere of Venus, trying to explain the greenhouse effect and so forth. An extraordinarily gifted publiciser, as you know, and a very sophisticated and broad person – a great person.

'We started together as friends, but we were in conflict a lot, because I from the beginning was very sceptical about life on Mars – and his view from the beginning was that wherever life can be, it will be.'

Mike Malin, whose extremely sensitive camera is currently in orbit around Mars on the Global Surveyor, remembers meeting Sagan as a student. 'Carl was a great guy. When I was a graduate student in the summer of 1971, Bruce Murray sent me to JPL to represent him in a meeting. I got there an hour before the meeting was due to start, so I sat patiently waiting. People started to filter in and they, sort of, looked at me but didn't make any comment.

'Eventually, some guy sat down next to me and said: "So who are you?" I said: "Well, my name's Mike Malin." He asked: "Do you work here in the lab?" I said: "No, I'm a new student of Bruce Murray's." "Oh, that's great, yeah," was his reaction. "Bruce is a really good guy." So I said: "Who are you?" He said: "I'm Carl Sagan." And I asked: "What do you do, Carl? Do you work in a lab?" You know, he said: "No – I'm a professor in astronomy at Cornell University." "That's cool," I said.'

Malin's other memory of Sagan reveals his compassionate side. 'My mother died just while I was writing my dissertation. So the dedication in my dissertation is to the memory of my mother Beatrice, who sacrificed her dreams for mine and who gave her life for mine. Carl Sagan is the only person ever to ask me the story behind my dedication, and we spent – oh, probably several hours talking about family and their importance and what my mother was like, and things like that.'

Chris Chyba, who now holds the Carl Sagan Chair at the SETI Institute in California's Silicon Valley, wrote to Sagan when he was studying theoretical physics at Cambridge. 'I noticed that I was spending most of my time in Heffers bookshop in the biology section, and in the rather thin origins of life section. And I thought that this was probably telling me something. I asked myself, how can I get involved in origins of life issues, given that I'm not a biologist – at least, not yet.

'The Carl I knew was a very warm and concerned person'

'I looked at who was doing what sort of research in the field, and it seemed that Carl Sagan's group at Cornell was directly addressing the question from the perspective of a variety of sciences. So I wrote to Dr Sagan, and it worked out very well. He encouraged me to come to Cornell and work with him as a graduate student.'

Chyba – a kind and compassionate person committed to issues such as human rights and nuclear non-proliferation – paints a glowing picture of Sagan. 'He was one of the smartest people I've ever known. I never heard him describe his memory as photographic, but that's how it seemed to me. He was a superb teacher and, above all, he was a quick thinker – he had

A Viking lander undergoes tests in the desert before flying to Mars. With its robot arm, weather sensors and TV camera, it was the most sophisticated probe of its day.

Lift-off of Viking 2 atop a Titan/Centaur launch vehicle on 9 September, 1975.

the ability to think on his feet extremely quickly to get to the heart of an issue. And very creative. He was a scientist who also made a decision that he would devote a lot of his time to other issues that were either in the field of education or in the realm of influencing policy.

'I think there's a broad perception that he changed as he grew older, and in fact Carl once remarked to me about the extent that he thought he had changed. And people assume that Carl somehow might not have been a warm individual, or there may have been powerful personality flaws. But the Carl I knew was a very warm and concerned person.

'I remember the time that my father was diagnosed with cancer, and I realised what a role Carl played in conversations with me. This was after I'd left Cornell, so I was talking to him as a friend at that point. Carl had very insightful, as well as sympathetic, things to say – that was characteristic of him. And after my father died, Carl made a point of finding a way to give me a break. He brought me back to Cornell to work with him on a project that was just intended as time to let me recover a bit. And that was completely Carl's own generosity.'

Carl Sagan and his biological allies got their own way on the new Mars mission in the end, and the Viking project went ahead on a grand scale. But not before a last-ditch attempt by Bruce Murray to scupper the plans. 'My role was to be the enemy. I tried to kill it,' he admits.

In the summer of 1975, two identical Viking probes blasted off from Cape Canaveral – atop Titan-Centaur launchers, instead of the originally mooted Saturn Vs. Each probe comprised two spacecraft: an orbiter to circle Mars and survey its features, and a lander to touch down gently on the surface. Viking 1 was scheduled to land in the Chryse Planitia region about 20° north of Mars's equator on the auspicious date of 4 July 1976 – the US bicentennial. But Mars refused to cooperate with the American dream. The landing site – which had been selected as a result of images from Mariner 9 – was anything but smooth, and much too dangerous to land on.

Mike Carr, head of the orbiter team, remembers vividly when the spacecraft's camera revealed the true extent of the site's roughness. 'It was really scary. We could see all this incredible detail. My God, I thought – we can't land there. There are so many obstacles in the field. So to me it was exhilarating to see that the camera was working properly, but my God, it's a landing site and we have this...'

'My God, I thought – we can't land there. There are so many obstacles in the field'

The team started an intense search for a new landing site. 'I had this army of people doing terrain assessments, you know, doing the geology,' recalls Carr. 'Every day, we'd get these new pictures coming in, and we'd have a meeting every day to assess the new data, presided over in this big room by the project manager.

'We'd put these wet prints up on the board and I had to interpret them – interpret the geology. And you know, I'd seen 'em only thirty seconds before, and tell everyone present what was there in the images. This went on for a few weeks, and we finally decided on the landing site we ultimately went to.'

Carr remembers running the two orbiters as the most intense period of his life. 'I think I had the most exciting job on Viking. The lander observations were restricted to what you could see – a few hundred yards. On the orbiter, we had access to the whole planet, and we got 55,000 pictures of Mars. We were looking at new territory all the time, and it went on for four years – it was just really exciting.'

In Mission Control at JPL, the science teams braced themselves for the first images from the Viking 1 lander – now safely on the surface 460 miles north-west of the original site, and some 16 days late in touching down. Would the technology work? 'It was tense,' recalls Mike Carr. 'But then there was the excitement of seeing the first pictures. Lots of new information was coming in, and the excitement was very real.'

Lou Friedmann, co-founder (with Bruce Murray) of The Planetary Society – a pressure group that would lobby the US government for continued Solar System exploration, following a post-Viking lull – was working at JPL at the time. 'I'll never forget that first panoramic shot of the Martian terrain,' he says wistfully. 'It still takes my breath away. It was the first view we had of an alien world – with all due respect to the Russians landing on Venus. It was alien and, on the other hand, it was terrestrial – like the American south-west, or the desert. You began to think that maybe there were similarities... maybe we could go to Mars to study ourselves.

'And these conflicting views – the alien world and the similar world – stayed with me. To me it represents the lure of Mars. We're exploring Mars to understand Mars, but we're also exploring Mars to understand ourselves.'

The lander revealed a dusty-pink world with salmon-pink skies – a result of dust particles blown into the thin Martian atmosphere. But there was nothing rosy about the Red Planet. It was bitterly cold, and almost airless. Drifts of fine Martian soil stretched for miles, as powdery as Antarctic snow (where some of the lander's tests had been carried out). Rocks and boulders of all shapes and sizes littered the scene. Many

'We're exploring Mars to understand Mars, but we're also exploring Mars to understand ourselves'

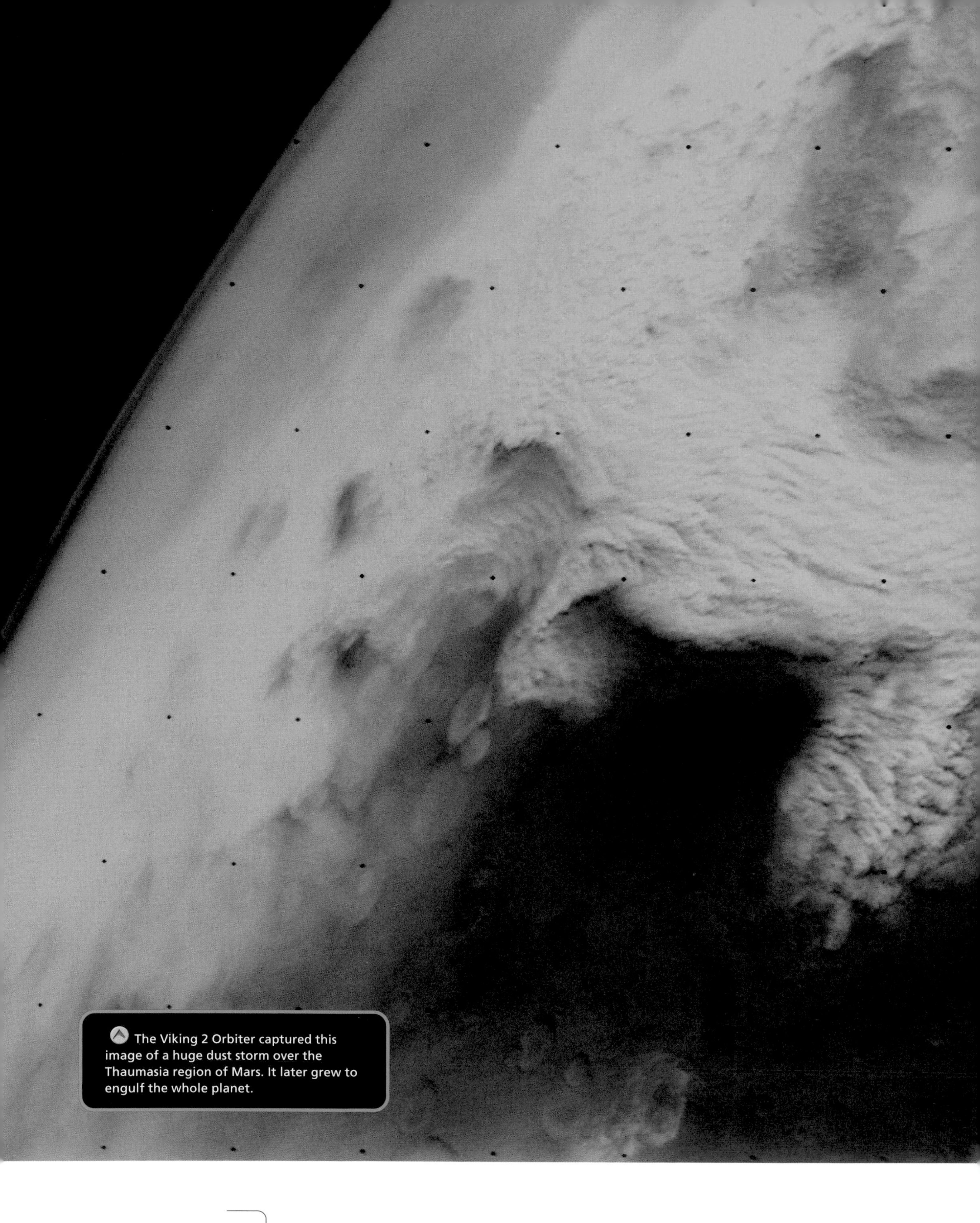

The Viking 2 Orbiter captured this image of a huge dust storm over the Thaumasia region of Mars. It later grew to engulf the whole planet.

The lander revealed a dusty-pink world with salmon-pink skies

were rough and volcanic in appearance, and some had small holes where gas had once bubbled through – like pumice.

The Viking 2 mission arrived at Mars a couple of months after its twin, and the lander touched down at a gently sloping site 50° north of the equator called Utopia. Both landers continued to perform flawlessly for many years. Each returned weekly weather reports, analyses of the Martian atmosphere, wind-speed readings, and thousands of pictures of the surface of Mars in all its moods. The Viking 1 lander lasted longest – a venerable six years – and after shutting down it was renamed the Thomas A. Mutch Memorial Station, after a brilliant young Viking team leader who was lost when climbing in the Himalayas.

As fate would have it, Bruce Murray – Viking's chief opponent – ended up being the man in charge of controlling the whole mission. 'Funny. The irony was that I was named director of JPL on 1 April 1976, and Viking actually landed on 20 July. So I was the person running it. I was the base commander, if you want. And I was opposed to spending $2 billion on Viking. I didn't think it was worth it.'

But despite its high price-tag, Viking had a great deal to offer. It was hugely inspirational. As Mike Malin points out: 'There was no Internet in those days, and it may have been harder to get information. But it was an era where essentially every rocket launched to a planet or the Moon, or with a human in it, was covered by all three networks. Viking was covered by all three networks live, and it landed at four or five in the morning. That was pretty good coverage for those days, I think.'

His younger colleague Ken Edgett leaps in like an excited puppy. 'I wasn't aware of Viking until a month before the landing. Then I got this magazine, and it said this thing was going to land on Mars. And I was like, wow – this is really cool, you know? I turned into a news junkie at that point. I had to get every bit of news I could get – magazines, newspapers, whatever.'

Steve Squyres believes that the Viking missions turned him from being a geologist into becoming a planetary scientist. 'In my third year

This panoramic view of Mars was the first picture taken by the Viking 1 Lander on 23 July, 1976. The large rock (centre) was dubbed 'Big Bertha' – but later renamed 'Big Joe' after howls of protests from feminists.

as an undergraduate, I had the opportunity to take a graduate-level course into the results of the Viking missions to Mars. I nearly got thrown out of the course because I was an undergraduate, but I convinced the professor to let me take it. Because it was a graduate-level course, we had to do some original research for a term paper. I remember that the professor was actually a member of the imaging science team for the Viking orbiter, and so he had all the images from the mission. Now, of course, nothing was digital in those days, so we just got big boxes of rolls of photographic prints. One afternoon I decided, oh, I'd better start thinking about my term paper. So I went over to the room where all the images were kept – it was called the Mars Room – and thought that I would sit down for fifteen or twenty minutes and flip through some pictures.

'I was in there for four hours – you know, sitting there and flipping through these pictures of this alien world. I suddenly realised that this was the kind of exploration I'd always wanted to do, and I went out of that room knowing exactly what I wanted to do with the rest of my life.'

Stunning though the Viking images undoubtedly were, it was the life experiments on the Viking landers that captured the imagination of the world. But, ironically, one scientist who had instrumentation on the landers was convinced from the outset that Mars had absolutely no life on its surface.

For a start, Jim Lovelock is not your usual kind of scientist. Although he once worked for Britain's Medical Research Council, he hankered to go it alone. 'It was an almost idyllic place to work at. It's just because I'm a cussed person that I wanted to go independent.' And that is what he has being doing for the past forty years, in an ancient mill house embosomed in the Devon countryside. 'I work from my own laboratory here, and fund myself by my wits.'

He is eighty-one but possessed of both boyish charm and mischievous sparkle by the bucket-load – not to mention a razor-sharp intellect. There is something about his whole being that is English through and through.

So how did a British independent scientist get to be working for NASA? 'They asked me if I'd join them on the first Surveyor mission, which was to go to the Moon, and analyse its surface. I think they asked me because I'd invented a number of sensitive detection devices, and they thought they might need to use them.'

At around the same time, Lovelock began working on the theory that would bring him international fame – the Gaia hypothesis. Our planet has evolved since its birth to suit living things, and life, in turn, has shaped our planet. Earth's 'oddities' – water, oxygen and life – are all interrelated. Lovelock thinks we should regard the whole planet as a vast living entity, which he calls 'Gaia' after the Greek earth-goddess.

This isn't a post-green, touchy-feely woolly notion. Gaia has come up with some clear guidelines that may help us in our search for life in the Universe. For instance, Lovelock has pointed out that the atmosphere of the Earth is very odd by cosmic standards – technically, it's in chemical disequilibrium. A fifth of it is oxygen – a very reactive gas that ought to combine with the iron in the rocks and the carbon in the plants at the drop of a hat. It hasn't: and that's because it's being continually topped up by expirations from plants. There are traces of methane in the air, which should similarly have disappeared by combining with the oxygen. It's there because of

'I went out of that room knowing exactly what I wanted to do with the rest of my life'

what comes out of the back ends of animals... In other words, chemical disequilibrium is a powerful indicator of life.

Long before Viking, Lovelock had predicted that we would find no life on Mars. 'If you go back to my *Nature* paper of 1965 – a hell of a long time ago – I said that the best way to look for life on Mars was to do an atmospheric analysis experiment. This experiment was done in France using the Pic du Midi Telescope, which looked at the infrared absorption of the gases in the Martian atmosphere. And this showed that the atmosphere of Mars was close to chemical equilibrium, and so the planet almost certainly did not have any life on its surface.'

Lovelock's involvement in the Viking missions was as an instrument scientist, rather than as a biologist. 'But I couldn't help being curious. In those days it was relatively small at JPL, and everybody talked with each other. And one of the space biologists asked me if I'd attend one or two of their meetings, which I did. I found out that nearly all the life experiments they were thinking of sending to Mars were so geocentric that I didn't think they'd have a dog's chance of working. And so I started criticising, and suggested they should look at the whole planet – instead of just at something on the surface.' But – as before – the biologists had the last word. And they were nothing if not ingenious. Each Viking lander carried a miniature laboratory, the size of a wastepaper basket, to perform the life-detection experiments on the surface by remote control. Nothing as sophisticated had ever flown to another world.

With its ten-foot-long arm, each Viking lander fed soil samples into its laboratory to be tested. There were four main experiments. Three looked at biological or chemical reactions with the soil, and one – the gas chromatograph mass spectrometer (GCMS) – broke down the soil into its basic atoms.

Two of the 'reaction' experiments – one which tested for plant life, and the other which looked for gas given off by cells – gave negative and inconclusive results respectively. Any reactions that took place, agreed the researchers, were down to chemistry rather than biology. But the third experiment had everyone sitting up and taking a considerable amount of notice. It was the 'labelled release' experiment of 'sanitary engineer' Gil Levin.

'We initially thought, my God, my God – there may be life there'

Levin explains his technique. 'It's very simple. The standard method of culturing micro-organisms is to put them in some kind of nutrient soup, and to wait several days until they start multiplying and you can see them. My technique simply added radioisotopes to those nutrient compounds. This meant that as soon as the micro-organisms started metabolising them, they would expire radioactive gas – which would be detected much more quickly than waiting around for a visible bubble. So the whole thing reduced about two days of waiting for evidence of life to about fifteen or thirty minutes.'

Mike Carr recalls the feeling in the Control Room when Levin's experiment yielded up copious quantities of radioactive gas. 'I mean, we initially thought, my God, my God – there may be life there. And then, of course, it all kind of waned.'

Ever since then, NASA's official line on the Labelled Release experiment is that it, too, discovered chemistry rather than biology. But Gil Levin absolutely refuses to take this assessment lying down – something which can irritate some of his colleagues. 'He's had his day in court,' says Lou Friedmann. 'He's been working on this for coming on to twenty-five

Viking Orbiter imagery has been computer-processed to reveal a dramatic view of the Valles Marineris, Mars's vast canyon system. It is over 2,500 miles long – ten times bigger than the Grand Canyon.

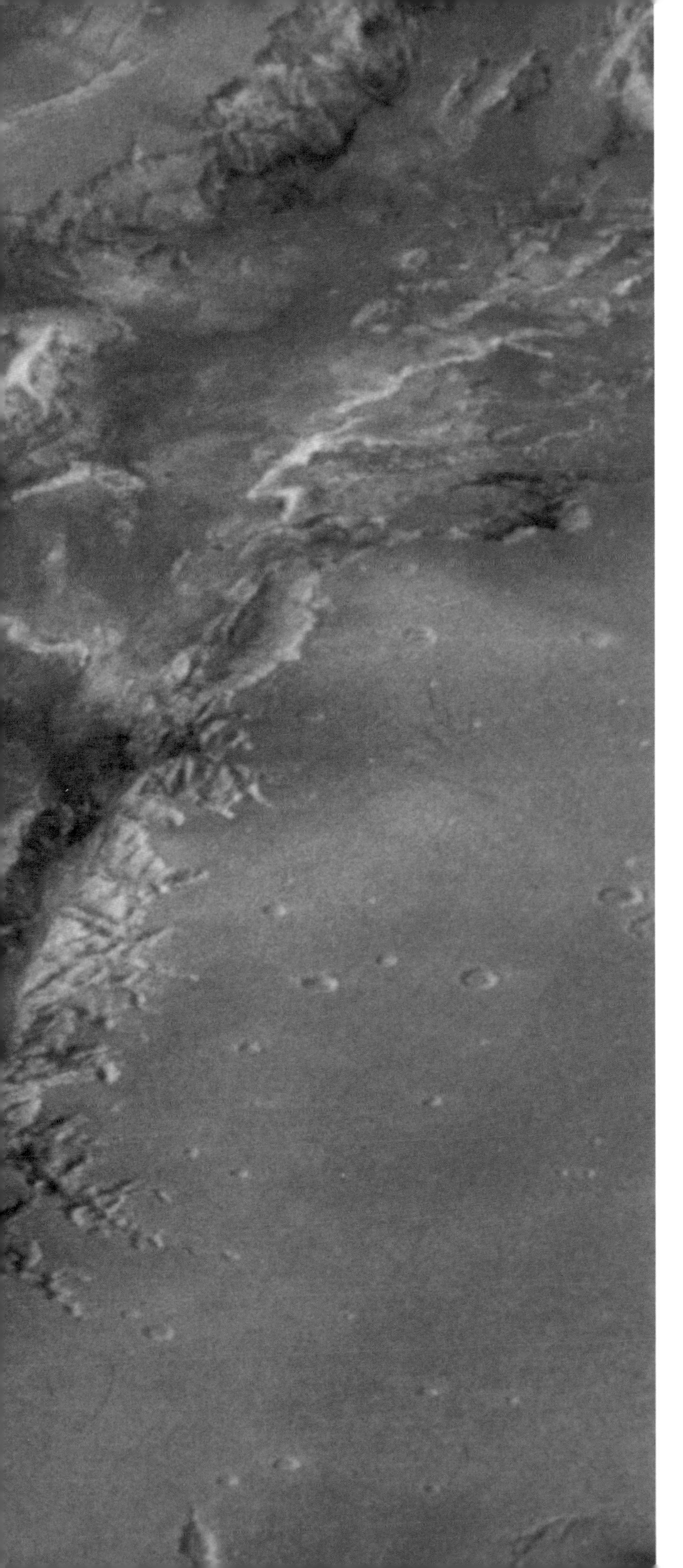

years, and he's not been able to convince the professional community.'

We decide that we'd like to talk to the guy to discover where the truth – if any – lies. It proves more difficult than we had anticipated. Finding the website for his company, Biospherics Incorporated, is no problem – except (as is also the case for Malin Space Science Systems) there are no contact details. So while we're in Washington we turn to the old-fashioned expedient of looking up the company in the telephone directory. Soon we are on the phone to a very guarded Gil Levin.

It turns out that he has recently appeared on television in connection with the life on Mars debate. He was promised a fair hearing. Instead, he found himself on a show surrounded by assorted loonies, nutters, Face-on-Mars freaks and flying saucer fiends. No wonder he is suspicious of the media. But he grants us an interview, and we head out in the direction of the Beltway.

In the conference room of his highly successful biotechnology and IT company, Gil Levin fills us in on some of the background. NASA, it appears, has always given Levin a rough ride. 'My problem is that I'm an engineer. I'm an engineer in a small company, and when my experiment began to work, NASA called me down and said: "We have a problem. It looks as though your experiment might be selected – it looks awful damned good – and you're just an engineer. What's more, you don't have a PhD – so how can you, if successful, go and talk to the National Academy of Sciences, go abroad to distinguished universities and report? So we want you to take on a senior investigator, and he will report it."

'I absolutely refused, and said: "I'm not going to give up this experiment. I'll go get myself a PhD." So I went to school and I worked – did both full-time for three years – took all the sciences and got my PhD in engineering.'

Sunset on Mars, as seen by Viking 1. It was a symbolic image: there were to be no further successful missions to Mars for another 21 years.

At first, everything seemed to bode well for Levin's experiment. He recalls the excitement in Mission Control at the time. 'When my experiment came up – it was about seven-thirty at night – the results clattered out of the computer, and we saw this curve and were amazed. We'd tested the Labelled Release experiment hundreds of times on Earth with thousands of different micro-organisms. We knew what those curves of response looked like, and here was one staring us in the face. We were astounded,

No experiment can be validated without controls, so Levin's team decided to 'kill off the bugs' in their sample by heating it to 160°. 'We waited, and finally the computer started spitting out again,' recounts Levin. 'There was zilch. A flat control line.'

'I'm not going to give up this experiment. I'll go get myself a PhD'

But Levin's problems were just about to start. Results from the GCMS experiment began to come in shortly afterwards. The gas chromatograph broke down the soil into its constituent atoms, and then analysed the make-up with its in-built mass spectrometer. It found many familiar chemical elements, including iron, silicon and oxygen. But there was absolutely no trace of carbon – the basic building block of life. How could Levin's experiment have detected life if there was no organic matter on Mars to make it?

> **There was absolutely no trace of carbon – the basic building block of life**

Worse was to come. As geologist Bruce Jakosky observes, Mars must be kept topped up by carbon from meteorites. 'You find organics on the Moon from meteorites. So something must be breaking apart the organics on Mars. The suggestion that's been pretty much accepted is that there must be oxidising agents, like hydrogen peroxide, in the Martian soil that would attack the organics and break them apart.'

The root cause of this weird chemistry is the lack of oxygen in Mars's atmosphere. It has a planet-wide ozone hole that allows harmful ultraviolet radiation from the Sun to penetrate all the way down to the surface. This would sterilise the soil, and also confer on it strange and unexpected chemical properties. So the consensus view coming out of the GCMS experiment is: not only are there no organics on Mars, but they are ripped apart as soon as they arrive. Not a healthy environment for making life.

We go over this ground with Gil Levin. He is warming to us. When he realises that we were once astrophysicists before going off to work in the media, he begins to feel he can trust us. He takes some papers out of a filing cabinet and places a computer print-out on the table. 'Here's the data. As you can see, it continued to build at a strong rate until the second day, but continued even beyond that for the full eight days of the programme. Here are the controls that were run, so that we have signals from terrestrial soils with living micro-organisms that bracket these curves.'

To our eyes, the Mars data does not look like a chemical reaction. Chemical activity builds rapidly and dies away. But the Martian curve builds steadily, and looks identical to the terrestrial controls.

Levin had one last-ditch attempt to convince NASA that he was on to something. He decided to heat samples of Mars soil within a narrow range of temperatures – the range of temperatures that kill off bacteria on the Earth. 'First of all, we showed that 51° definitely destroyed the signal. But, secondly, we showed that 46° didn't destroy it – it inhibited it by 30 per cent. And that's just the way that in the laboratory here we distinguish *E. coli* from the rest of the coliforms, because *E. coli* can survive beyond 37°, while the others cannot.'

It was all to no avail. NASA had turned its collective back on Levin. 'It was political,' he acknowledges. 'They had to come down with a decision, and they hate to retract a decision. If you go to people from NASA and you say, "What do you think of the Labelled Release experiment?" they'll say, "Oh, that's garbage, you know. Levin keeps saying the same thing over and over again."

'In 1986 I was asked to speak at the tenth anniversary of Viking, and what I did was list, oh, maybe fifteen possible explanations of a non-biological nature for the Labelled Release experiment results – and I showed errors in each one of those. And for the very first time, I said: "It's my opinion that it's more probable than not that the Labelled Release experiment detected life on Mars." There was an uproar. Since then I've been essentially ostracised by the NASA-supported community.

To our eyes, the Mars data does not look like a chemical reaction

'Obviously, the finding of life on Mars is one of science's biggest grab-backs, and nobody wants to surrender that very easily – especially when there are missions yet to come, and experimenters anxious to perform. I think that brass ring is going to be held out until the point is made so overwhelmingly that life *has* been detected.'

Two experimenters anxious to perform are young chemists Danny Glavin and Oliver Botta, who work at the Scripps Institution of Oceanography in La Jolla, California. It's our first visit here. Since we were students, this place has been a legend to us – a combination of the romance of exploring the deep oceans, plus its wild location on the rugged southern California coast. But the Institution of our dreams turns out to be a tired, grey 1960s concrete building.

This doesn't seem to worry Glavin and Botta, who are happily rushing in and out of their office doing timed tests on equipment in the laboratory across the corridor. Their field is life-detection – and now they hope to take their techniques, tried and tested on Earth, to Mars. But will they do any better than the Vikings? 'We have to remember the Vikings were in 1976,' points out Glavin. 'But we've recently done some really interesting experiments with this new Mars Organic Detector, the MOD. We're building this instrument with some folks up at JPL, and hopefully it'll fly in 2005.

'What we've shown with the MOD is that you can detect amino acids with very high sensitivities. And using the detection limits of what the Viking GCMS measured for these compounds, and comparing it with our instrument, we estimate that Viking would have missed on the order of 30 million bacteria cells per gramme of soil.

'So there could have been cells in the soil, but Viking wouldn't have seen them. The MOD is on the order of a hundred times more sensitive. So we're really excited about the chances of finding something on the surface of Mars.'

Oliver Botta reminds us how much our knowledge about the tenacity of life in extreme environments has advanced since Viking. 'The problem with the biology package on Viking was that it was really designed to look for terrestrial analogues of life – which was not a very broad view of how life could have looked on Mars. And we also have to realise that, in the 1970s, we didn't know of this variety of life forms we have on Earth – high temperature, low temperature, and high acid, or whatever.'

Extremophile expert Jonathan Trent of NASA's Ames can even contemplate life in an environment of hydrogen peroxide. Although he's sceptical that any life forms can exist on the surface of Mars, because of the lack of water, he's already experimenting with the effects of peroxide on organisms. 'They have wonderful adaptations for getting rid of peroxide. For example, there's an enzyme called catalase. When you have a wound and put peroxide on to sterilise it, the peroxide will bubble. The catalase is transforming the peroxide into water and oxygen, and the bubbles are actually oxygen coming off. Many organisms produce catalase, and it isn't inconceivable that they could cope with the peroxide levels on the surface of Mars.'

JPL's Pam Conrad also broadly agrees that there is no life on the surface of Mars today, because its atmosphere is in equilibrium –

'We estimate that Viking would have missed on the order of 30 million bacteria cells per gramme of soil'

as pointed out by Jim Lovelock. 'But chemical equilibrium is time-dependent,' she observes. 'If you take your snapshot at the right time, you might see a disequilibrium. In the case of Mars, if we're looking at the wrong time in Martian history and life has been and gone, we may find fossil evidence. We would certainly like to walk up to a creature and say hello, but any life – extant or not – would be phenomenal. It would be a huge change for the way that we look at our role in the Universe.'

So where does the jury stand nowadays on the Viking life findings?

David Wynn-Williams, a geologist at the British Antarctic Survey, says: 'When Gil Levin evaluated his system in the Antarctic, he was able to show biological activity at very low densities of micro-organisms. So he's partly right – the GCMS was not as sensitive as his system – but I need more evidence to be convinced that he did actually see microbiological activity on Mars.'

'Mr Mars', Ames's Chris McKay, adds: 'If Gil's experiment was the only experiment we had, its results would be consistent with a biological source. But it was inconsistent with the other experiments. Explanations relying on a chemical reaction with the soil are more plausible, but I should say that none of these explanations has been proven.'

Bruce Jakosky looks at the question with the benefit of hindsight. 'We gave Mars our best shot, but that was over twenty years ago. There are lots of opportunities for life to have existed which wouldn't have been detected, and – as a result – we still do not have an answer to that question.'

NASA's John Rummel remembers how Viking's non-detection of life cast a shadow over the Agency. 'When I first got to Headquarters with the odour of what was assumed to be a negative result on Viking, there were many astrophysicists quite willing to prove to me that life couldn't exist anywhere in the Universe, including the Earth – and at days in Washington, I agreed with them.'

So what is Rummel's verdict on Viking?

'Not proven.'

Meanwhile, Gil Levin bides his time. He is heartened by the 'not proven' verdict. But NASA still stonewalls him, and a number of scientists doubt his competence. Does he let it get to him? 'I'm not resentful,' he smiles. 'I'm kind of amused. You know, no matter what they do, my data are not going to go away.

'And I do believe that it will be established one day, and I'm sure that when it is established, whoever does it will take the credit. But none the less the original data are there.'

FEAR AND PANIC!

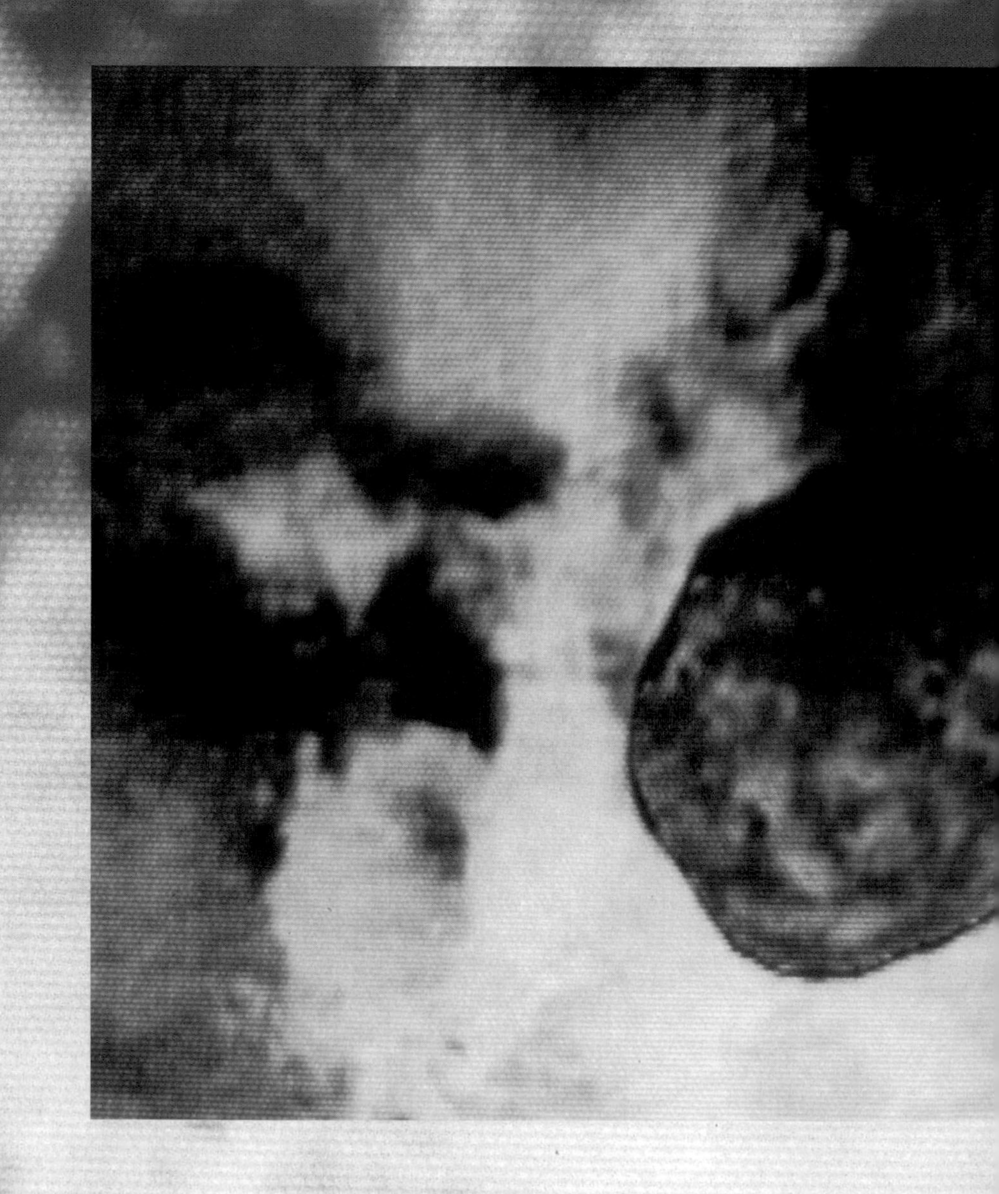

CHAPTER 6

'Mercury and Venus have no moons; Earth has just one moon; then there's Mars; and then you have Jupiter that has four moons,' Dick explains. 'Because of this progression, Kepler thought that Mars should have two moons.'

The miniature world of Lilliput is the most famous of the strange lands visited by Jonathan Swift's fictional traveller Lemuel Gulliver in his satirical novel *Gulliver's Travels*. But it was far from the only one. After his adventures among the diminutive Lilliputians, Gulliver travelled to the land of the giants, Brobdingnag. And his third voyage took Gulliver to Laputa, an island floating in the air.

Coppin del. et lith.

J'adressai à cette lune nouvelle et inconnue tous les signes qui me parurent pouvoir m'y faire prendre en pitié.

Lemuel Gulliver spots the flying island of Laputa, in a French edition of Jonathan Swift's classic novel. It was the home of a race obsessed with astronomy, who had made prescient observations of Mars.

Perhaps because they hover hundreds of feet in the air, the Laputians are particularly concerned with astronomy. They worry that the Earth may be swallowed by the Sun, that a comet may burn the world to ashes or that the Sun itself may be extinguished. 'They are so perpetually alarmed with the apprehensions of these, and the like impending dangers, that they [cannot] sleep quietly in their beds... When they meet an acquaintance in the morning, the first question is about the Sun's health.'

Mars does have two moons – but they were not to be discovered for another 150 years. How could Swift be so prescient?

The Laputian astronomers have telescopes far more powerful than any in Europe. They have observed 93 comets and catalogued 10,000 stars. But the most intriguing comment is Gulliver's account of their observations of the Red Planet: 'They have likewise discovered two lesser stars, or satellites, which revolve around Mars.'

Mars does have two moons – but they were not to be discovered for another 150 years. How could Swift be so prescient?

The mystery deepens when we look at the details. 'The innermost is distant from the centre of the primary planet exactly three of his diameters, and the outermost five.' The inner of Mars's two real moons is 2.8 Mars-diameters out, and the other lies at a distance of 6.9 Mars-diameters.

Perhaps most remarkable is Swift's statement that 'the former revolves in the space of ten hours'. All the moons known at that time revolved around their parent planets at a fairly sedate rate, much slower than the rapidly spinning planet beneath. With Mars spinning round in twenty-four and a half hours, Swift was proposing the first moon to move more quickly than its planet. Lo and behold, the inner moon of Mars *does* turn out to have a period much shorter than Mars's day.

'The discovery of the moons of Mars...is really one of the great adventure stories in the history of astronomy'

Is this evidence that Swift was psychic – or had been informed by visiting extraterrestrials? Astronomical historian Steve Dick thinks not. 'There's a lot of interesting prehistory to the discovery of the moons of Mars,' he says.

Dick starts with the German astronomer Johannes Kepler, who first worked out how the planets move around the Sun. At the same time – this was the very early seventeenth century – astronomers had turned the first telescopes to the sky, and discovered four moons going around Jupiter.

'Mercury and Venus have no moons; Earth has just one moon; then there's Mars; and then you have Jupiter that has four moons,' Dick explains. 'Because of this progression, Kepler thought that Mars should have two moons.' And because no one had seen these moons, they would have to be small and close to the planet itself.

Swift wasn't the only author to follow Kepler's lead. In 1752 the French author Voltaire described beings from the Dog Star, Sirius, who make an excursion to our Solar System. Reaching Mars, 'they saw two moons which wait upon this planet and which have escaped the gaze of astronomers.'

Not that the astronomers were being altogether idle. In the late eighteenth century, Britain's leading planetary astronomer, William Herschel, searched for any moons orbiting Mars – long, hard and, ultimately, unsuccessfully.

The discovery had to wait almost another century. And to follow up this quest, we have arrived at the US Naval Observatory near Washington DC. The neo-classical main building is set in the exact centre of a circle of parkland, to keep down any electrical interference from passing traffic. Part of the park is taken up with the official residence of the US Vice-President – and it's a running wry joke with the observatory's astronomers that very few VPs have ever bothered to wander across the lawns to view the splendours of the heavens.

Steve Dick leads us along a dark-panelled corridor to the imposing circular library. Dick is the observatory's official historian, and is also in charge of the International Astronomical Union's committee on the history of astronomy. 'The discovery of the moons of Mars, here at the Naval Observatory, is really one of the great adventure stories in the history of astronomy,' he begins. To emphasise its importance, he points out a highly personal commendation framed on the wall, from Abraham Lincoln to the moons' discoverer, Asaph Hall.

Hall was in charge of the observatory's biggest telescope, with a lens twenty-six inches in diameter. 'It was the largest telescope in the world when it was built in 1873,' Dick proudly declares, 'and it was still almost the largest refractor [lens-telescope] in the world in 1877 when he made that discovery.'

At that time, the observatory wasn't at its present salubrious location on the heights near the fashionable suburb of Georgetown.

Built in 1873, the US Naval Observatory's Great Equatorial Refracting Telescope has a lens 26 inches in diameter at its top end; it is still one of the largest refractors (lens-telescopes) in the world.

Eagle-eyed Asaph Hall tracked down Mars's two moons with the 26-inch refractor. Fellow astronomers later and unfairly doubted his skill, as he couldn't see the (now discredited) 'canals' of Mars!

It was down by the Potomac River – at the appropriately named Foggy Bottom. And 1877 was a momentous year in our relationship with the Red Planet. Mars was making an unusually close approach to Earth, and all over the world astronomers eagerly anticipated the opportunity to scrutinise the planet in unprecedented detail. In Italy, Giovanni

Glorious news from the skies – Brilliant Discovery with the Great Telescope – Two satellites of Mars found by Prof. Hall

Schiaparelli caught the first glimpse of the soon-to-be-notorious 'canals'. Here in Washington, Hall had other thoughts on his mind.

Perusing the astronomical textbooks – as opposed to the fictional works by Swift and Voltaire – he was struck by the repeated statement 'Mars has no moons'. Equipped with one of the world's most powerful telescopes in this auspicious year for Mars, he was determined to check this out, once and for all.

At the beginning of August, Hall began searching for faint 'stars' near Mars, which could be tiny and hitherto overlooked moons. Night after night any possible culprits turned out to be real stars that happened to lie behind Mars. A lesser astronomer might have given up – or an astronomer with a lesser wife.

'Hall was a persevering person,' Dick continues, 'but not as persevering as his wife, as it turns out. His wife, Angeline Stickney, was his math teacher before he became an astronomer.'

Angeline Stickney was a formidable figure in an age when few women had careers, let alone careers in the sciences. As her student, Asaph Hall and his friends would try to devise questions that she could not solve: but to no avail. According to the custom of the time, though, she gave up her job when she married Hall and became an unpaid assistant at the observatory.

'She and the entire family – they had four or five sons – supported him in his search,' says Dick. 'Hall had searched many nights for the moons of Mars and had seen nothing and was about to give up. And his wife said, as she sent him off after dinner to the telescope: "Try one more time." That was the night he found them.'

What Hall glimpsed was a tiny speck of light very close to Mars. He jotted down its position just minutes before the rising fog from the Potomac blotted out the sky. After a few cloudy nights, he spotted the tiny moon again. And as he watched one further night, to fully confirm the discovery, Hall discovered a second moon – slightly brighter, and even closer in to Mars itself.

'It was a tremendous observation,' enthuses Dick, 'because we now know that they are extremely small. And it was an interesting psychological thing too.' After Herschel's intensive hunt, most astronomers had decided Mars had no moons, so they had given up searching. 'But Hall decided that he would look – and found them. And the rest is history.'

The headlines and accolades were international. The *Washington Evening Star* exclaimed: 'Glorious news from the skies – Brilliant Discovery with the Great Telescope – Two satellites of Mars found by Prof. Hall.' France's greatest astronomer, Urbain Leverrier, called it 'one of the most important discoveries of modern astronomy'. It was one of the few occasions when an American Vice-President felt impelled to make the short trip to the observatory's great telescope.

The new moons, of course, needed names. Reading about the discovery at Eton College in England, the science master Henry Madan recalled verses from Homer's great epic on the siege of Troy, the *Iliad*. As Ares – the Greek version of the war-god Mars – prepared to take part in the struggle, he called to his two attendants Phobos and Deimos, representing fear and flight. 'He spake, and summoned Fear and Flight to yoke/ His steeds, and put his glorious armour on.'

What names could be more appropriate for the moons that attend the planet Mars?

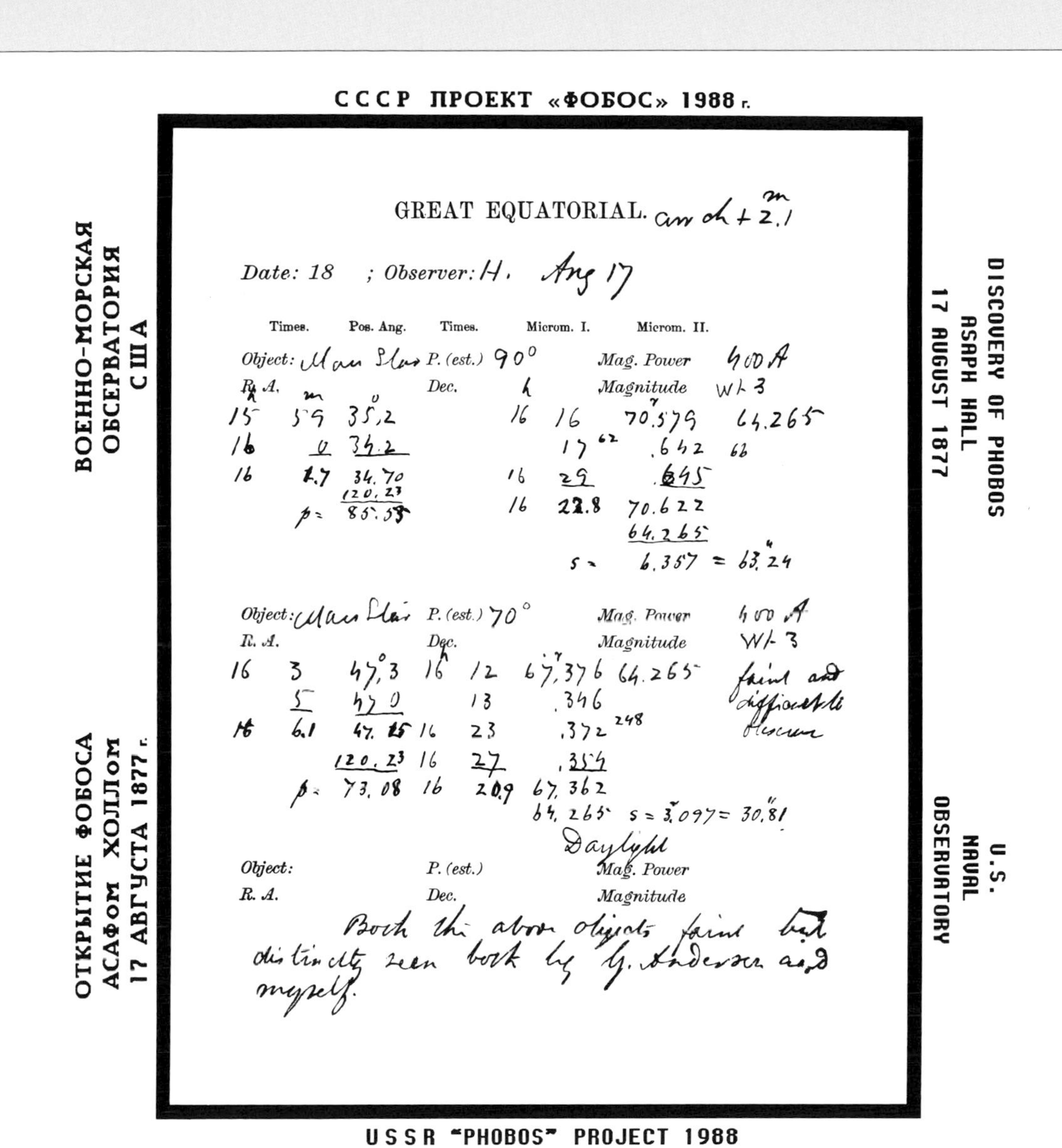
СССР ПРОЕКТ «ФОБОС» 1988 г.

ВОЕННО-МОРСКАЯ ОБСЕРВАТОРИЯ США

DISCOVERY OF PHOBOS
ASAPH HALL
17 AUGUST 1877

GREAT EQUATORIAL. an ch + 2.1

Date: 18 ; Observer: H. Aug 17

Times. Pos. Ang. Times. Microm. I. Microm. II.

Object: Mars Star P. (est.) 90° Mag. Power 400 A
R. A. Dec. h Magnitude W/- 3
15 59 35,2 16 16 70.579 64.265
16 0 34.2 17 .642
16 1.7 34.70 16 29 .645
120.23
p = 85.53 16 22.8 70.622
64.265
s = 6.357 = 63.24

Object: Mars Star P. (est.) 70° Mag. Power 400 A
R. A. Dec. Magnitude W/- 3
16 3 47.3 16 12 67.376 64.265 faint and difficult to measure
5 47.0 13 .346
16 6.1 47.15 16 23 .372
120.23 16 27 .354
p = 73.08 16 20.9 67.362
64.265 s = 3.097 = 30.81
Daylight

Object: P. (est.) Mag. Power
R. A. Dec. Magnitude

Both the above objects faint but distinctly seen both by G. Anderson and myself.

ОТКРЫТИЕ ФОБОСА
АСАФом ХОЛЛом
17 АВГУСТА 1877 г.

U.S. NAVAL OBSERVATORY

USSR "PHOBOS" PROJECT 1988

A bilingual plaque attached to the Soviet Phobos missions reproduces Asaph Hall's discovery observation of Phobos, while checking previous sightings of Deimos.

The inner and bigger moon became Phobos ('Fear'), while the outer moon took on the role of Deimos (now generally translated as 'Panic').

There was, incidentally, to be an interesting sequel to this naming of the Martian moons. When a planet was discovered beyond Neptune in 1930 – at another American observatory – the name 'Pluto' was suggested by a niece of the same Henry Madan, eleven-year-old Venetia Burney.

For decades after the discovery, astronomers knew very little more about the tiny Martian moons. But poring over the path of Phobos as it endlessly circled Mars, a Soviet scientist suggested that something extremely odd was going on. It heralded a long-lasting Russian fascination with these tiny worlds.

A moon made of solid rock shouldn't feel much drag from the atmosphere...Shklovskii concluded that Phobos must have a very low density...In other words, it would be an extraterrestrial spacecraft

In the forefront of the Soviet initiative was Roald Sagdeev, head of Russia's unmanned space programme during its glory years. To track him down, we don't need to travel far from the US Naval Observatory. Sagdeev is now a professor at the University of Maryland, on the outer fringes of the suburbs surrounding Washington.

Sagdeev seats us in a sparsely furnished meeting room, where English and Russian titles vie for space on the bookshelves. He's quick to point out that he's not just a professor of astronomy, but also director of the East-West Space Science Center.

'When I first came here, there was no indication that the Soviet Union would have a difficulty with the space programme – or even that it would be dismantled at all. I left with the tacit permission of Gorbachev. The original idea was that I would be spending fifty per cent of my time in Russia, fifty per cent here, trying to help.'

Sagdeev is now permanently in the US. 'If I went back to Moscow, I would just be adding to the army of unemployed Russian scientists.' And what's happened to his other former colleagues? 'Some of them are scattered in the world, like the debris from a big bang.'

Sagdeev began his scientific life as a nuclear physicist. After the launch of Sputnik 1 – the first satellite – in 1957, he moved towards studying hot gases and radiation in space, and before he knew it he was appointed director of the Space Research Institute. 'It was like a kind of shock therapy!'

Sagdeev had a crash course in learning about the planets. Among the more mainstream science, he came across an intriguing idea from his elder colleague, Iosef Shklovskii. 'He was a very original thinker,' Sagdeev recalls, 'and not everything he was suggesting eventually proved to be true – but he had really tremendous imagination.'

Over the decades since Hall's discovery of the moons, astronomers had found that Phobos is very gradually spiralling downwards, towards the surface of Mars. That's not entirely a surprise. Phobos is so low down that it may be brushing through the top of the planet's atmosphere, and friction could be dragging it downwards – the fate that met the Earth-orbiting Mir space station, which spiralled downwards and crashed into the Pacific Ocean in March 2001.

'But Shklovskii saw that Phobos's loss of altitude is enormously high,' Sagdeev continues. A moon made of solid rock shouldn't feel much drag from the atmosphere; it should just force its way through. Shklovskii concluded that Phobos must have a very low density. 'His interpretation for Phobos being so light is that it is an artificial object, like an empty shell. In other words, it would be an extraterrestrial spacecraft.'

The first Soviet probes to Mars failed to show whether Shklovskii was right or wrong; they failed en route. But in 1969 cameras on the American Mariner 7 revealed Phobos was not a metallic artificial structure; it was an irregular lump of dark rock. And to put the idea finally to rest, astronomers have concluded that Phobos isn't descending at anything like the rate that had worried Shklovskii.

After Mariner 7, other American robot eyes scrutinised Phobos and Deimos in intimate detail. But the Martian moons were still very much on

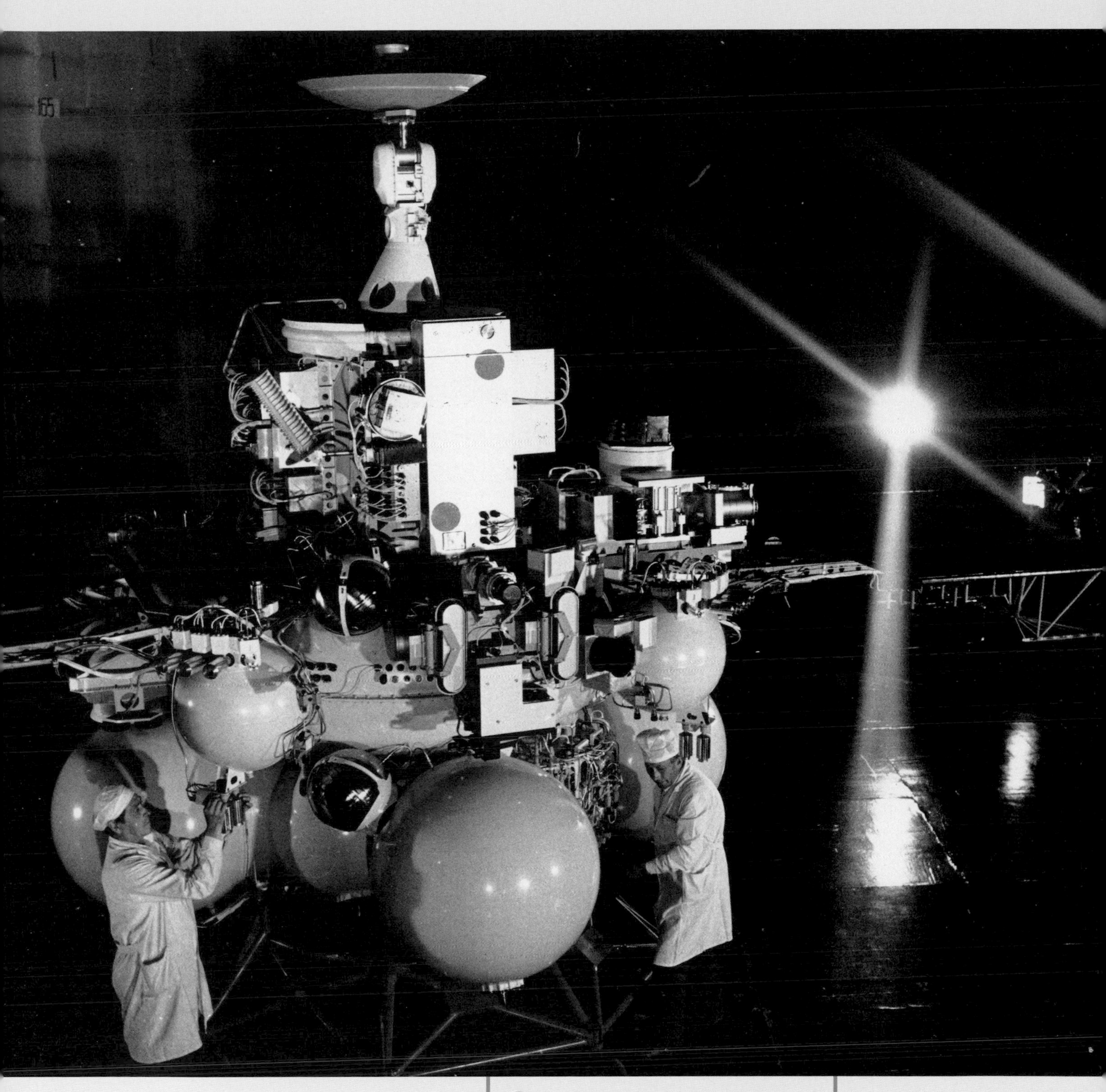

Built in true Soviet heavy-engineering style, the Phobos 1 probe bristles with instruments to investigate the Martian moon at close quarters.

December 1954 · 35 Cents

Astounding
SCIENCE FICTION

Mars seen from Phobos

From Phobos, Mars looms in the sky some 30-times bigger than the Earth looks from our Moon. On this 1950s magazine cover Martian canals are still clearly in vogue!

Sagdeev's mind. In 1988 his agency launched a pair of spacecraft towards the Red Planet. The names of the probes – Phobos 1 and Phobos 2 – show their intended destination. And the mission was nothing if not ambitious.

'The idea was that the spacecraft would gradually change its orbit and come very close to the Martian moon Phobos,' Sagdeev explains. 'It would fly very close to the surface, at an altitude of 30 to 50 metres – like a cruise missile.'

Cameras on board would reveal the moon's surface in almost embarrassing detail – Sagdeev predicted they could pick out individual grains of sand. And the spacecraft would also discover what Phobos was made of, by firing a small laser at Phobos's surface and analysing the cloud of vaporised rock. 'We were going to use Star Wars technology,' says Sagdeev with a twinkle, 'though we were not planning to knock Phobos down!'

And they would also drop tiny landers on to the moon's surface. 'Inside the company, we called them "micro-Vikings",' Sagdeev reminisces. 'One of them would look almost like Viking, with a small camera and a compartment for chemical experiments. A second lander had the capability to make several jumps, like a grasshopper.'

In the event, one Phobos probe failed on the long interplanetary flight; the second failed as it approached Mars's moon. They were two of the many casualties strewn along the path of Mars explorations, victims of 'the Curse of Mars' (see Chapter 7).

But these failures haven't quashed the Russian spirit for exploring Mars's moons. 'After the failure of the first Phobos project,' admits Sagdeev, 'I developed a new driver, probably somewhat similar to the subconscious drive of criminals to revisit the scene of their crime!'

And Sagdeev's old colleague, Mikhail Marov, still on his home soil at Moscow's Keldysh Institute of Applied Mathematics, has caught the bug. 'We already have some experience because of the 1988 missions. So we've found our own niche, Phobos, where we can concentrate our efforts.'

We catch up with Marov at the International Astronomical Union's meeting in Manchester, England, where he's hoping to drum up international support for the new mission.

'We call it the Phobos Sample Return,' he enthuses, 'and hope it will go sometime between 2004 and 2007.' A massive spacecraft will head from Earth-orbit to an orbit around the Red Planet. Like Russia's 1988 missions, the Phobos Sample Return mission will nuzzle up to the larger of the Martian moons and drop a lander. This 'micro-Viking' will drill into Phobos to extract the eponymous sample. A smaller rocket will shoot the sample back to Earth.

A massive spacecraft will head from Earth-orbit to an orbit around the Red Planet

'So we start with a seven-tonne spacecraft,' grins Marov, 'and the capsule that returns to Earth is only 12 kilograms – and within that just 100 grams of the real soil from Phobos.' But scientists such as Sagdeev see that soil as worth far more than its weight in gold. 'If I had a second ticket,' he says, 'I would fly such a small spacecraft to Phobos.'

Until this mission arrives, our best views of Phobos and Deimos have come from those workhorses of Martian exploration, the two Viking orbiters of 1976. At that time, one

A striking celestial alignment of Phobos (foreground), Deimos and the Sun appears above Mars's giant volcanoes and canyons in this artist's impression of the view from a manned Mars spacecraft.

surprised scientist blurted out 'Phobos looks like a diseased potato' – a description that has yet to be surpassed!

Both moons are certainly potato-shaped. And with lengths of seventeen miles for Phobos and ten miles for Deimos, they are mere specks in the cosmic ocean. Even the larger moon is no bigger than the Isle of Wight or Nantucket Island. On a motorway, you could easily drive from one end to the other in quarter of an hour.

'Phobos and Deimos are comparable in size to the smaller asteroids,' says Clark Chapman, an expert on these minor worlds that orbit the Sun between Mars and Jupiter. The asteroids are fragments of a planet that never formed. And from his office in Boulder, Colorado, Chapman has recently been making the first close-up study of an asteroid through the eyes of an orbiting spaceprobe. This small world, Eros, is also a cosmic potato, just a little bigger than Phobos.

Does this mean that the Martian 'moons' are really just asteroids that have fallen under Mars's gravitational spell and have been bound into orbit around the Red Planet?

'It's not that easy to capture asteroids into orbit around Mars,' Chapman replies. 'Basically, things in interplanetary space just coast along, and something coming close to Mars will swing around and go away again. What you really need is to have an asteroid approach Mars very slowly and then have something slow it down further, right when it's next to Mars.'

The passing asteroid might have hit a moon already orbiting Mars – but that just begs the question of where the first moon came from. Or, continues Chapman, 'maybe Mars once had a much thicker atmosphere and an asteroid came and skimmed the top of the atmosphere and slowed down. That's probably the best-favoured idea, but it's not very convincing either.'

Chapman puts up an alternative idea. 'Conceivably, these moons actually were made of material from Mars itself, when some kind of a cratering collision sent a bunch of debris into space and some of it bumped around and ended in orbit.' In that case, Phobos and Deimos would have had a similar traumatic birth to the famous Martian meteorite containing possible 'fossilised bugs'. It was splashed even further out and eventually collided with Earth.

But Chapman is the first to admit that this idea isn't totally convincing. In particular the moons are much darker than the rocks of Mars, so they don't seem to be made of the same material as the planet. 'It's really a bit of a mystery why those moons are there,' Chapman muses. 'It's not obvious how they formed, or where they came from.'

Phobos's dark grey colour is in stark contrast to the bright ochre surface of Mars in this view from the Soviet Phobos mission – one of the few Soviet/Russian colour images of Mars and its moons.

Astronomers like Chapman are wont to refer to both moons in a single breath. But apart from their potato shapes and dark colour, they are by no means identical twins. The photographs taken by the Vikings reveal that Phobos is rough and cratered, while its smaller companion is uncannily smooth.

Deimos is draped in a thick layer of dust that smooths out whatever contours may lie below, like snowdrifts smoothing out a winter's landscape on Earth. Two hollows in the moon's curves hint at craters below, named Swift and Voltaire. Chapman suggests an origin for these dusty drifts: 'If there's a cratering impact that sends debris shooting away from Deimos, it's captured into orbit around Mars – and then re–accumulates back on to the surface of Deimos.'

'It's really a bit of a mystery why those moons are there,' Chapman muses. 'It's not obvious how they formed, or where they came from'

While the Moon has one-sixth of the Earth's gravitational pull, Phobos's gravity is only one-thousandth of what we're used to on Earth. Jump too energetically, and you will launch yourself into space!

After such a collision, the debris spread around Deimos's orbit will form a temporary ring around Mars – making the Red Planet a miniature version of the Lord of the Rings, the planet Saturn.

'On Phobos we see large cracks and crater chains,' Chapman continues, 'and an enormous crater that probably reflects an impact that came close to breaking Phobos into pieces.' The huge crater is named Stickney, after the wife whose perseverance led to the discovery of the Martian moons. Asaph Hall himself is commemorated by a smaller crater on Phobos.

Lesser of the twin moons, Deimos is only nine miles long. It is blanketed in deep drifts of dust, which may swirl out into space to form a temporary dusty ring around the Red Planet.

The terrific impact that blasted out the Stickney crater probably shattered the interior of Phobos, leaving it riddled with gaping voids. This would explain the long cracks running from end to end of the moon. The apparent crater chains, Chapman surmises, could be 'places where the dusty loose surface soil is collapsing and draining into these cracks'.

This raises the tantalising possibility that future explorers may find tunnels into Phobos, where they could shelter from the dangers of solar flares and the other hazards of space. Not quite Shklovskii's giant alien spaceship, but something of a natural alternative!

Chapman certainly looks forward to the human exploration of Phobos and Deimos. 'I think they might be very interesting outposts from which to sit and observe Mars' – though he counts himself out of any Mars exploration himself. 'When I was a young kid, I used to think about going to Margaritifer Sinus on Mars, but seeing the rigours of life on Mir and the International Space Station, I think that's more for people who enjoy being cold and hungry and in pain – so I'll let other people do that!'

John Rummel, NASA's Planetary Protection Officer, backs up the idea. 'Phobos and Deimos are some of the best places you can imagine going to.' For a start, it takes a lot less energy to come and go from the moons than having to descend to the planet's surface, and then fight its gravity to get back up to space.

'And they provide a great platform from which to supervise the robotic exploration of Mars,' Rummel continues. 'You get rid of the delays that arise in radio communication from Mars to Earth, and you have access to satellites you can put into Mars orbit. So these are great missions to envision.'

In 1989 NASA commissioned a team to study how astronauts might explore Phobos. It will be a very different expedition from the Apollo exploration of our own Moon. Gravity is the main difference. While the Moon has one-sixth of the

A huge crater (right) on Phobos is named Stickney, after discoverer Asaph Hall's wife. The large rubble-filled trenches suggest the moon's interior may be fractured, with huge caverns inside.

Earth's gravitational pull, Phobos's gravity is only one-thousandth of what we're used to on Earth. Jump too energetically, and you will launch yourself into space!

Instead of thinking of Apollo, the report suggests: 'For the crew, the experience might be likened to working on a large, dusty and unequipped spacecraft.' If they do want to make use of Phobos's weak gravitational tug, the astronauts will have to wear a massive 'turtle shell' to increase their weight. The NASA team suggests giving each astronaut a weight of fifteen pounds on Phobos, which means wearing a shell that would weigh five tons on Earth!

For the far future, the ultimate space pundit Sir Arthur C. Clarke predicts: 'Phobos and Deimos will have a very important role to play in the exploration of Mars – though I hope they don't hack them for propellant material.' Sir Arthur's concern arises from suggestions that the Martian moons may contain organic molecules and ice, which could be mined to manufacture rocket fuel.

But Clarke himself made a more destructive suggestion in his science fiction novel *Fountains of Paradise*. 'I suggested taking them apart, and building them into an elevator down to the surface of Mars,' he recalls with a wry laugh.

If our spacefaring descendants do refrain from dismantling Phobos, it will ultimately smash down on to the Martian surface naturally, between 100 million and 1,000 million years in the future. All that will remain is one more fresh crater pockmarking the Red Planet.

In the meantime, Deimos is slowly spiralling outwards. Eventually, it will break free from Mars's gravity and spin free to circle the Sun as a minor planet in its own right.

But, in the meantime, Mars's pair of moons represent a challenge to astronomers' understanding of the Solar System. 'Phobos and Deimos really are unique objects,' Clark Chapman concludes. 'They are mysteries – and we'll find out just how interesting they are as we explore Mars closer.'

THE CURSE OF MARS

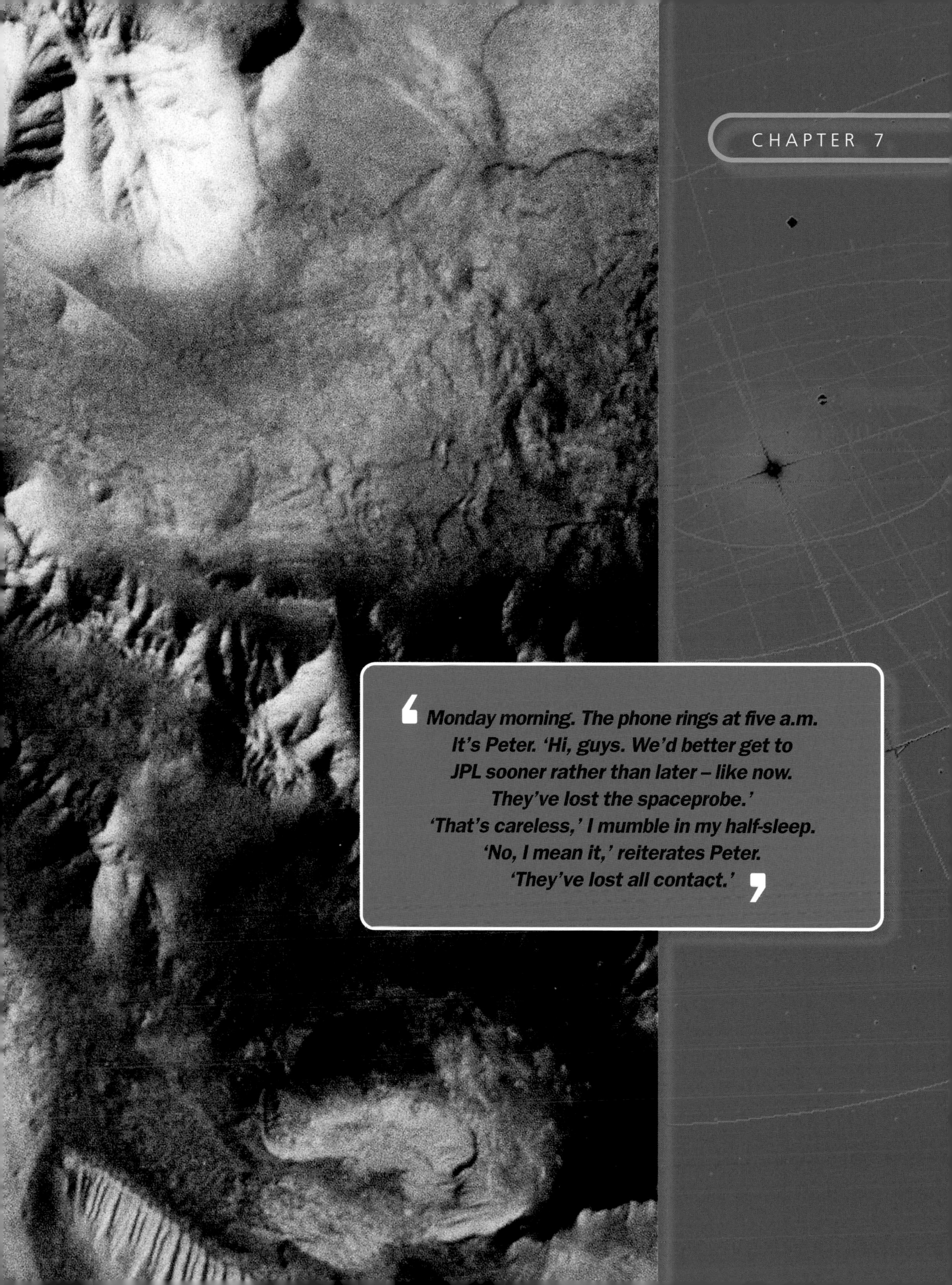

CHAPTER 7

'Monday morning. The phone rings at five a.m. It's Peter. 'Hi, guys. We'd better get to JPL sooner rather than later – like now. They've lost the spaceprobe.'
'That's careless,' I mumble in my half-sleep.
'No, I mean it,' reiterates Peter. 'They've lost all contact.''

It was one of those flights from hell. Sunday 22 August 1993, and we'd been forced into booking too late to get a direct flight to Los Angeles. So we cross the Atlantic sandwiched in the middle of Economy – it's a fallacy that the media travel Business – surrounded by gaggles of package-tour passengers on their first trip to the States. A couple of beers at Newark Airport, and we take to the skies again in a plane that's in no hurry to get to LA. We touch down at seven-fifteen in the evening, still having to pick up a hire car and negotiate the web of freeways that will eventually lead us to Pasadena.

Artist's impression of the doomed Mars Observer spacecraft. The $900 million probe was one of the last expensive NASA missions launched: afterwards the Agency adopted its 'faster, better, cheaper' policy.

But we figure that it's all worth it – because this weekend marks NASA's return to Mars, seventeen years after the Viking missions. Eleven months earlier, the agency had launched a massive $900 million spaceprobe designed to go into orbit around the Red Planet and scrutinise it as never before. The Mars Observer would create a global portrait of the planet, and in the NASA hype of its glossy brochure: '... it will also help prepare the way for future explorations – missions of the 21st century that will carry automated rovers and, eventually, astronauts to the Red Planet.'

We can't let the opportunity pass, and persuade Channel 4 television in the UK to commission a programme about the spaceprobe and its significance. Channel 4 won't commit to a full budget but – at the eleventh hour – give us enough money to fly to the States for a few days to film the arrival of the Mars Observer at the Red Planet. Hence the flight from hell...

Our destination is the Jet Propulsion Laboratory (JPL) in Pasadena – a leafy and beautiful Spanish-inspired city north of Los Angeles. JPL, tucked under the misty foothills of the San Gabriel Mountains, controls most of the missions to the planets – most famously, the Voyager probes to the outer Solar System – and the Mars Observer is its latest baby.

More to the point, our destination that night is the Holiday Inn next to Pasadena's sprawling Convention Center. We sigh with relief as we pull up at the familiar, ziggurat-shaped building. It's not a luxury hotel in any sense of the word, but it's a kind of emblem for the thousands of film crews, journalists and scientists who come to do business at JPL. Peter McPherson, who's doubling

[The Mars Observer] marks NASA's return to Mars, seventeen years after the Viking missions

Nestling under California's San Gabriel Mountains, the Jet Propulsion Laboratory is the world's nerve-centre for missions to the planets.

as both director and cameraman on this shoot, is there already. So it's drinks at the bar and a chance to discuss what we're going to film tomorrow.

Monday morning. The phone rings at five a.m. It's Peter. 'Hi, guys. We'd better get to JPL sooner rather than later – like now. They've lost the spaceprobe.' 'That's careless,' I mumble in my half-sleep. 'No, I mean it,' reiterates Peter. 'They've lost all contact.'

This was not the first time – and will certainly not be the last – that probes bound for Mars have gone missing. Ever since missions began, Mars seems to have exerted a peculiar curse on our spacecraft. The first people to find out were the Russians.

On 1 November 1962 the then Soviet Union launched Mars 1 towards the Red Planet. Sixty-six million miles into the journey, radio contact was lost – and was never re-established. 'That was the first attempt to send a man-made object in the direction of Mars,' muses Roald Sagdeev, former Director of the Space Research Institute in Moscow.

Mars 2 and Mars 3, which followed in 1971, reached the planet but failed to send back data. In fact, the Mars 2 landing craft crashed (although, technically speaking, it *was* the first object from Earth to land on another planet). Two years later, Sagdeev was invited to head up the Soviet unmanned space programme. 'When I took my chair as director, I learned that in three months' time we were going to launch four heavy unmanned spacecraft to Mars – Mars 4, 5, 6, 7, if I didn't mix the numbers. Each of them was about five thousand kilos and required a dedicated Proton launcher. Imagine... during a narrow period of two weeks... four Protons launched from Baikonur!'

These were mighty beasts indeed. Although not as powerful as the famous Saturn V rocket which put men on the Moon, they still had more brute force than anything else in the American rocket fleet. The prospect of seeing four of them launched in such a narrow window must have seemed like something out of space fantasy.

This was not the first time – and will certainly not be the last – that probes bound for Mars have gone missing

Except that Sagdeev knew better. Weeks into his new job, he found that the rockets' Mars-bound cargoes were flawed. 'I discovered that the latest tests of the systems of these spacecraft indicated that some microcomponents or transistors were doomed to fail. Technological defects were discovered.'

This came as no surprise to veteran Soviet space-watcher and self-confessed space nut Jim

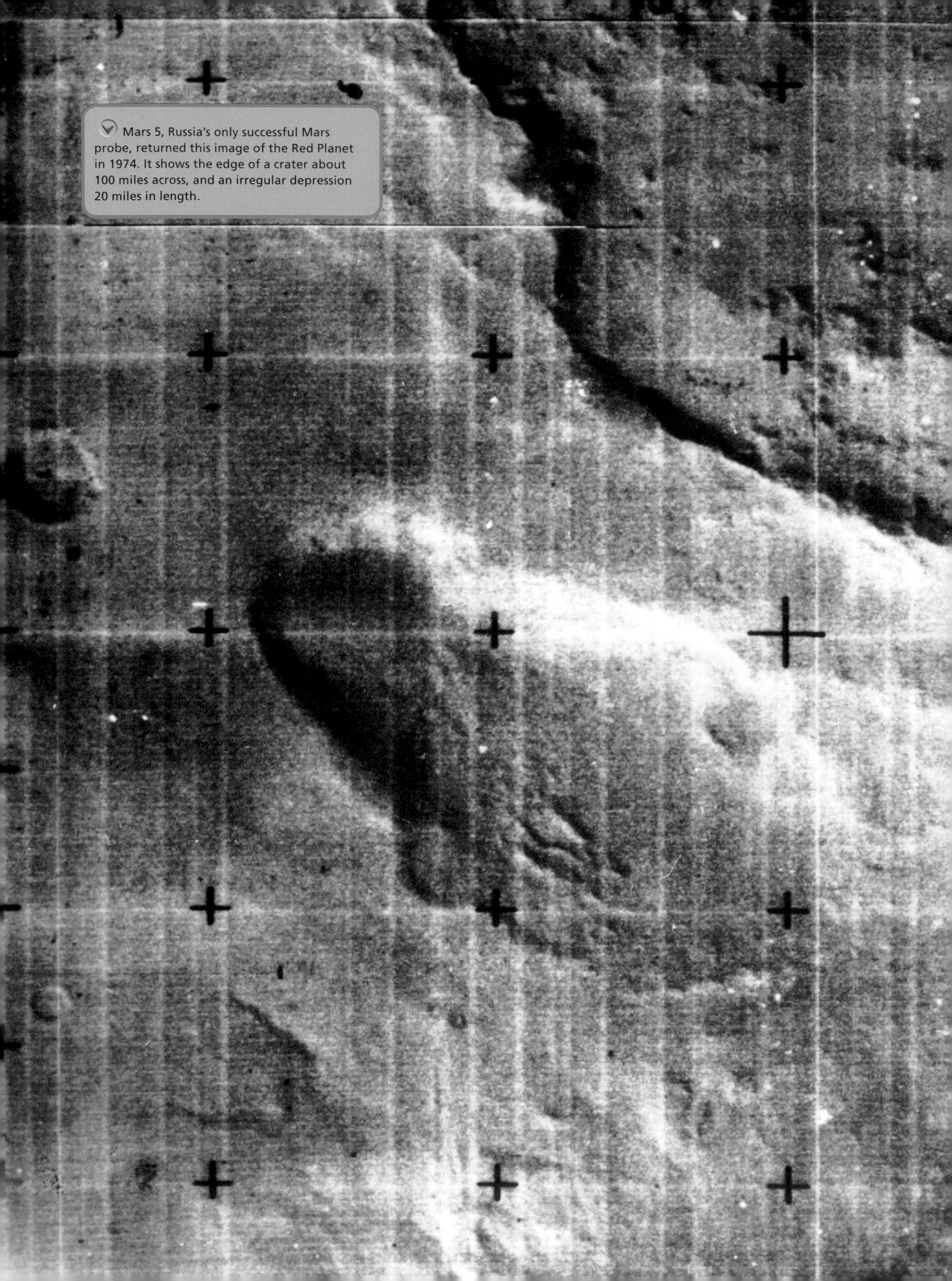

Mars 5, Russia's only successful Mars probe, returned this image of the Red Planet in 1974. It shows the edge of a crater about 100 miles across, and an irregular depression 20 miles in length.

Oberg. A former NASA engineer – with a well-known antipathy to the Agency – Oberg has spent decades probing behind the scenes of the intensely secret Soviet space programme. His classic book, *Red Star in Orbit,* contains grainy photographs of cosmonauts which have been clearly retouched to obliterate personnel who had been killed, or removed from the space programme for other reasons.

We meet Oberg at our hotel opposite the NASA complex in Houston and we travel in his car to an all-you-can-eat Mexican buffet.

'The key technology that's different between Russia and America – the weakness you could put your finger on – would be the lifetime of electronics. Russian electronics tend to be much shorter-lived in space than Western-made electronics.'

'The government and we all knew that we were gambling – it was like Russian roulette. Russian Martian roulette'

But back in 1973 Roald Sagdeev's problem was that defects in the electronics were discovered *before* the launch of their four Mars probes. 'There was a huge discussion whether to stop the programme,' he recalls. 'But the inertia was so strong, so much money was already invested, and also, you know, most of the hardware was already integrated deep inside these huge spacecraft.

'So finally the decision was to fly – no chance to replace, too short a time. The government and we all knew that we were gambling – it was like Russian roulette. Russian Martian roulette, in this case.

'So what happened in the end was that, out of the four spacecraft, two did not reach Mars. And the particular transistor at fault was also implemented in different systems – the on-board computer, the scientific instruments – and statistically there is a probability that these were decaying.'

To Jim Oberg, it was also a question of the Soviet philosophy – brawn, not brain. Although their microelectronics might not work, their launchers did. 'Their launch costs were so low that they could launch numerous replacement satellites into Earth orbit. But when you start moving out beyond the Earth and out into interplanetary space, this kind of technology approach, this kind of philosophy, became impossible to overcome.'

In the event, Mars 5 – which reached the planet in 1974 – became the only Russian probe that has had any success at Mars. As British planetary researcher Colin Pillinger tartly puts it: 'Out of twelve missions, they've had eleven failures – and one that produced some data, but not much.'

Mars 5 was an orbiter, and it sent back a number of images to Earth. Roald Sagdeev recalls the extraordinarily primitive state of Soviet technology at the time. 'There was a photo camera, and the films were processed on board, with all the necessary chemicals. Now you can do this unmanned photo-processing in every shopping mall – you know, if you pay a few bucks you can get it. But to imagine that it was carried inside a Soviet-made spacecraft in 1974 to orbit Mars – it was a little miracle. You got film and then had to scan it, then the data were coded and sent by radio to the Earth.

'Out of twelve missions, they've had eleven failures – and one that produced some data, but not much'

'And I think some of these pictures became quite prominent in the press. But they were, of course, much inferior compared to Viking.'

But Sagdeev was to change all this – including the brawn versus brain philosophy. Under his visionary leadership, the Soviet space programme slowly pulled itself up by its bootstraps. After several successful missions to Venus, and a close encounter with Halley's Comet, the Soviets felt ready to dice with Mars again. This time, it was an ambitious mission to land on one of Mars's tiny moons – Phobos. In mid-1988 two Phobos probes took off from Earth bound for Mars. Just a couple of months into the mission, the controllers had to send a message to the on-board computer on Phobos 1 telling it that there would be no communication from Earth during the late-summer holiday period. Roald Sagdeev takes up the story: 'Apparently, this coded message had a digital error. Instead of saying something sensible, the message was read by the on-board computer as a signal to close the caps of the small nozzles which were used to control the attitude of the spacecraft.' But the controllers were oblivious of the error. 'They happily left for a few days – it was the end of August, when kids have to be sent back to school again, and they were hurrying home to their families. So a few days later, when they got back, they couldn't get any signal from the spacecraft. Most likely it was already frozen after losing orientation. The batteries would have completely dried up, so there was no more source of energy.'

Phobos 2 fared marginally better. It actually reached its target. But as Sagdeev sadly reflects: 'It approached Phobos to the distance of about a hundred kilometres and started to send the first images. At this moment in the process of operations, we lost the spacecraft – it went silent.'

Trying to find out what happened was not helped by the political regime at the time. Roald Sagdeev remembers: 'It was Gorbachev, but still Soviet time. And the aerospace industry did everything to produce a cover-up. What was known is that the last signal was very weak. So the engineers concluded that the antenna dish was turned away from the direction of the Earth.' Sagdeev assumes that a computer glitch in some way caused the failure. 'Once the computer malfunctioned, the spacecraft's orientation would go wrong. Gradually, it would lose its energy juices, because the solar panels would be turned away from the Sun... and that's it.'

The next Mars probe was also a spectacular failure. And this time it wasn't Soviet or Russian. The Curse of Mars hit NASA's Mars Observer.

Pasadena, Monday 23 August 1993: it's a beautiful morning. The smog hasn't built up yet, and the San Gabriel Mountains are sharply outlined against the clear, blue sky. As we dash into JPL to find out what's happened to the missing spaceprobe, we glimpse the radio masts surrounding the venerable Mount Wilson Observatory, 6,000 feet up. As kids, we were brought up on the exploits of its ancient 100-inch telescope. Is seeing it a good omen?

of absurdly young computer scientists brainstorming over what might have happened to the probe. But even though there is no signal, the intended 'arrival at Mars' tomorrow is still scheduled to go ahead. On Tuesday lunchtime a signal will be sent towards the spacecraft commanding it to slow down by firing two of its thrusters, and allow itself to be captured by Mars's gravity.

Next day, there's still no signal. But there's still hope among the scientists that the planned orbit insertion might work some magic.

One-forty p.m., Pacific Daylight Time. A signal from the Deep Space Network of radio telescopes in the nearby Mojave Desert commands the Mars Observer to fire its thrusters. We – and thousands of JPL employees – collectively hold our breath. The probe stays utterly silent. That night, we get very, very drunk.

We decide to stay on one more day to see how the mystery unravels. Arriving at the gates of JPL, we have to push through an unruly rabble of protesters. They are accusing NASA of having deliberately destroyed the probe.

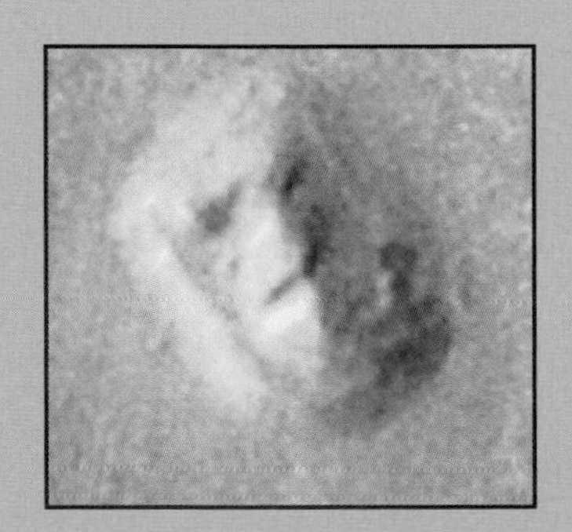

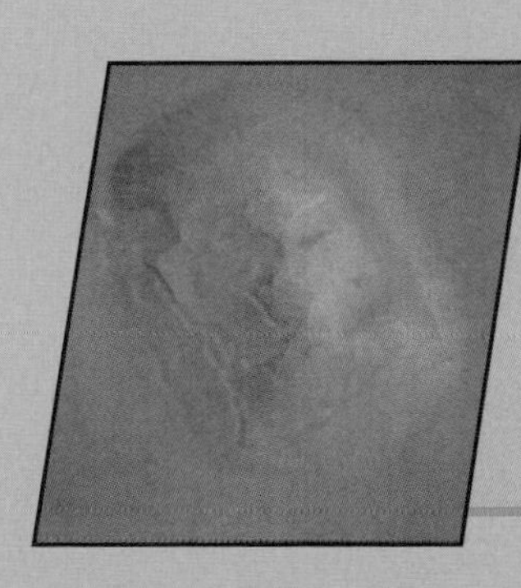

(left) Viking captured this image of the controversial 'Face on Mars' in 1976. But the latest image from the Mars Global Surveyor (right), acquired on 3 April 2001, reveals The Face as a complex, dissected mesa

It seems not. At JPL there's a mood of cautious optimism, but there isn't the usual confidence you feel about the place. We interview a few of the scientists working on the mission, including the chief project scientist, Arden Albee. He seems bewildered.

Here and there on the campus are pockets of intense activity. Huddled in small rooms are groups

The ringleader is holding aloft a poster which shows a feature on the Martian surface. It looks remarkably like the head of a man wearing a helmet. The 'Face on Mars', they claim, is a gigantic artefact of a Martian civilisation, like the Egyptian pyramids. The protesters' argument is that NASA knew the Mars Observer was scheduled to image the area where The Face is situated, and couldn't bring themselves to admit that there might be intelligent life – or had been life – on Mars. So NASA blew up the probe.

Peter, our director/cameraman, is fascinated, but we can't even bring ourselves to go near the protesters. So we leave Peter to conduct an interview with them as we go in search of any news about the probe. There's none. We seek out Arden Albee and ask how he's feeling. 'The loss was a shock for everyone,' he says despondently. 'You know – I was asked by a local reporter what did I do? And I held up my hands and showed her that I'd been out pulling ivy in my back yard, and my hands were all bloody from pulling ivy. I guess it's a shock... you know, it's hard to react.'

We fly back the next day, reflecting that NASA has lost its probe and that we've lost our TV programme.

Seven years later and we're back in the States still searching for answers as to the nature of The Face, and to find out what really did happen to the Mars Observer. Our travels take us to the offices of Malin Space Science Systems in San Diego, run by the brilliant and highly individual Mike Malin. He rarely gives interviews, but his right-hand man, Ken Edgett, was brought up on one of our books. Malin acquiesces.

'We knew that opening those valves was going to be among the most risky activities the spacecraft conducted'

Mike Malin is the world's leading expert on space imaging, and he designed the camera for the doomed Mars Observer. Of all the scientists who worked on the mission, he was one of the few who was not surprised when the probe was lost. 'There's a story about that, about which I don't think the public is well aware. We knew there was a risk in this mission before the launch in February 1992. In East Windsor, New Jersey – where the spacecraft was being built – it was determined that the valves and the propulsion system had some finite potential of failing.

'And the problem I think is one that derived originally from a philosophy that you should simply take an Earth-orbiting spacecraft and send it to Mars. But propulsion systems on satellites in Earth-orbit do everything they're going to do in the first few days – whereas a propulsion system going to Mars has to wait eleven months before it does some things.

'If the valves were opened as they normally would have been opened in Earth-orbit, within a few days of launch, everything would have been OK. But there was more or less a 100 per cent probability that the helium pressurant – which is used to force the fuel and oxidiser together – would leak out in the course of the months to Mars. And so you had, basically, a 100 per cent probability of not being able to do the Mars-orbit insertion manoeuvre.

'The basic story is that we knew that opening those valves was going to be among the most risky activities the spacecraft conducted.'

Eleven months later, Malin's worst fears were confirmed. Three days before the planned orbit insertion on 24 August 1993, the craft was due to pressurise its on-board propellant tanks – which necessitated opening the valves. Malin recalls the occasion vividly: 'I came into JPL that day – which was a Saturday – and tried to get on our voice-control line to listen in. I couldn't, so I called the project manager who was in his office with his predecessor and the head of the JPL Mars programme. They were listening to their voice-box that was relaying the activities, and they put me on speaker-phone so I could listen with them while this went on.'

Before firing the valve, they had to shut down the radio transmitter. Amazingly, in these days of microelectronics, the radio included a travelling wave tube amplifier – a vacuum tube with electrodes inside it – which performs more reliably in certain situations. The controllers were worried that firing the valves could shatter the tube, and so took no chances.

'So they turned off the travelling wave tube amplifier, and the spacecraft was now out of communication. A certain time later the valve was supposed to fire – this was a little after five in the afternoon – and then by about quarter to six the spacecraft was supposed to have turned the tube amplifier back on and re-acquired the signal by around six in the evening. And we never heard from it.

'And I basically went through my reaction of having lost the spacecraft at that point. I didn't know the spacecraft was going to fail, but I was prepared for it.'

Malin recalls his side of the day when we arrived at JPL with hope still in our breasts. 'I went up there on Tuesday for the insertion. Every one of the company drove up to participate in this – even though we knew it was gone by then.'

Ken Edgett, Malin's colleague, was then a graduate student at Arizona State University. His memories of the event are positively spooky. 'I was working for a guy called Phil Christensen, who was also a principal investigator on Mars Observer. We had just had a teacher workshop on the Saturday, and that ended at five. And as Malin pointed out, these events occurred between five and six p.m. that afternoon.

'And I don't remember specifically thinking that it was gone that evening, but I called a friend of mine and said, "You know, there's something wrong with the Observer?" It was like he thought I'd told him that his father had died. I mean – it was really heavy. This friend of mine was just "no way". You know: "This kind of thing doesn't happen in this business."

'On Tuesday, NASA went ahead and did an event as if it were going into orbit, and it was all on NASA TV. We were holding an open house at the university for kids and families to come in and participate with us. And Phil Christensen gathered us together first before we did the public event. I don't remember exactly what he said, but I remember his voice cracked. And then I said: "That's it, it's gone, because he knows it's gone." It was hard on us. People were really sad and crying.'

Even now, there's no one explanation for the loss of the Mars Observer. The most probable scenario is that the leaking helium pushed some of the oxidant into the fuel line. When the fuel tanks were pressured up, the fuel and oxidant mixed, causing the fuel line to burst. The destabilised spacecraft then went into a spin, and all contact was lost.

But what about the conspiracy theory that NASA blew it up to avoid having to face up to The Face on Mars? 'One of the fallacies about modern American life is that there are conspiracies, OK?' observes Mike Malin. 'But there are no conspiracies here.'

Nevertheless, Malin acknowledges that there's a widespread fascination about The Face. 'I don't understand it, but it appeals to a certain attitude in the public at large. It's mysterious... it allows you to use your imagination. It allows Mars to be an imaginary place.'

The Face was first imaged by one of the Viking Orbiters in the summer of 1976. It's a huge structure a mile long, almost a mile wide, and nearly 1,500 feet high. It quickly leaped to fame as a result of the attentions of Richard Hoagland, a former museum space science curator. Hoagland, who has an almost evangelical

'It was hard on us. People were really sad and crying'

The 'sandworm', or 'glass tunnel' on Mars. Is it an alien artefact? Or is it just a trick of the light – a trough filled with sand dunes seen under strange illumination?

presence, pointed out that it looked artificial: like an artefact of a civilisation. He began to publicise his convictions through a series of personal appearances that were more like religious rallies than scientific presentations. Very soon, he was surrounded by an entourage of 'disciples', all keen to disseminate his word.

The Face is located on a flat plain on Mars called Cydonia in Mars's northern hemisphere. Hoagland has pointed out that there are a number of other formations on Cydonia that also look artificial. Within a radius of about twelve miles of The Face, there are several pyramidal objects with sharp, angular sides. Hoagland calls the whole area 'The City'.

Until 1998, Hoagland and his disciples had only the low-resolution Viking images to support their cause. Enter Mike Malin. Although Malin's camera was lost with the doomed Mars Observer, he designed a very similar camera that would fly on the highly successful Mars Global Surveyor – which is still orbiting Mars and returning data.

Malin, a geologist by background, had clear ideas of the targets on Mars that he wanted to image with the Global Surveyor. But NASA had other ideas. They wanted high-resolution images of The Face. Malin dug his heels in. 'I said, "NASA did not select me as principal investigator of an experiment to delegate the authority to tell me what to spend my scientific resources on." The project manager respected my opinion, but felt that his prerogative was to direct me to do what he wanted me to do.

'I ultimately received communications from Headquarters directing me to do what the project manager was telling me to...sacrifice the science for The Face'

'So I asked NASA Headquarters for guidance on this. After a considerable amount of delay, I ultimately received communications from Headquarters directing me to do what the project manager was telling me to do, and sacrifice the science for The Face.

'I'm so sick of the thing, and it has poisoned my relationship with JPL, and the project manager, and the earlier project manager for Mars Global Surveyor.' In the event, Malin – with his arm twisted halfway up his back – secured a high-resolution image of The Face shortly after midnight on 5 April 1998. But at least it showed the object in its true colours. 'It's kind of like a really big butte in Monument Valley,' observes Ken Edgett, 'but there are some uniquely Martian processes that have modified the slopes on the thing. It takes us into: "What is the Global Surveyor finding?" There's a lot of alien geology – a lot of terrain that's not like the Earth.'

So alien that it may have convinced the world's greatest science visionary that Mars *could* be inhabited. Although we suspect that Sir Arthur C. Clarke could be talking with his tongue firmly wedged in his cheek.

'I've no doubt there's life on Mars,' he volunteers, halfway through a conversation about something completely different. 'Have you seen those incredible images? I can't imagine what the hell is going on. I mean, there's a sandworm and there's that forest at the south pole, and I haven't seen them in any newspaper anywhere at all.'

The rest of the dialogue goes like this:

HC: 'The sandworm – what does it look like?'

ACC: 'You've not seen it? It was a glass tunnel thing you see, you must have seen that?'

HC: 'No, I haven't. I tried to find a website for it.'

ACC: 'Well, this is absolutely insane. I mean, everyone would go crazy over it.'

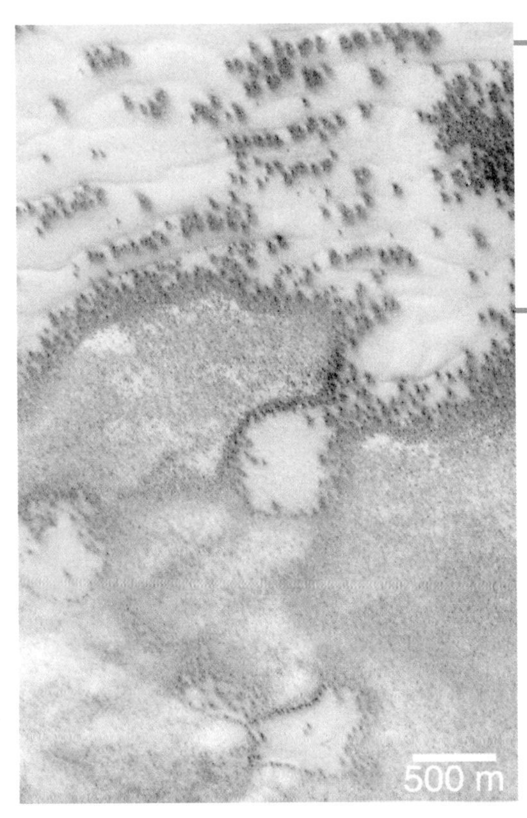

'The forest is absolutely convincing,' maintains Arthur C. Clarke. 'A lot of trees in a snowy landscape.' Actually, the 'trees' are dunes defrosting in the spring sunshine.

The Mars Climate Orbiter was meant to be a Martian weather satellite, which would also search for water. But it succumbed to the Curse of Mars...

Another failed probe: the Mars Polar Lander, scheduled to arrive less than three months after the Climate Orbiter, ended up crashing to Mars's surface.

HC: 'How big is it?'

ACC: 'It's a glass tunnel, miles long.'

HC: 'Oh, right.'

ACC: 'The forest is absolutely convincing. A lot of trees in a snowy landscape.'

HC: 'What? Gosh.'

ACC: 'I mean, what's going on here? Let me check my mailbox, Space Miscellaneous. I'll send you the images.'

Ken Edgett sighs with exasperation. 'Is he still pushing that thing around? He and Richard Hoagland have been doing it for almost a year now, it seems. I get e-mail on it about once a week. I usually ignore it.'

So – what is the sandworm?

'It's a trough. The features on the trough are dunes. We see these things – narrow troughs with dunes and ripples in them – all over the planet, and we saw them in Viking high-resolution images too. So there's nothing new here. For reasons that escape me, those that see a "tube" are inverting the feature in their minds.'

With the successes of the Mars Global Surveyor, and that of the 1997 Pathfinder mission on the surface of Mars, it seemed that NASA was having a honeymoon period with the Red Planet. An ambitious future was planned: flotillas of spaceprobes, robotic rovers, darts that would penetrate deep into the surface, even Martian planes flying through its thin atmosphere. But then the Curse of Mars struck again...

After the demise of the expensive Mars Observer probe, the tough and effective NASA chief Dan Goldin had instituted a new policy: 'Faster, better, cheaper.' External contractors vied with one another to cut costs yet build efficient spacecraft. NASA was learning from its expensive mistakes – or so it thought.

Two 'faster, better, cheaper' probes were launched towards Mars in 1998. Together, the Mars Climate Orbiter and the Mars Polar Lander came in at only one-third of the cost of the Mars Observer. 'We were very excited,' recalls project scientist Richard Zurek when we talked to him in his office at JPL. 'We thought we had a good combination, because we were doing two things. We had a weather satellite that was the Mars Climate Orbiter, which would tell us about the weather as it occurs, and it would do it for a full Mars year. We'd be looking at water, where it comes from, and when it appears in the atmosphere during the year.

'And we were also going to go with that water cycle on the ground, with the Mars Polar Lander. It would be the first time that any of our craft had landed in the polar regions. And we were going to search for the ancient climate. By digging down through the beautiful

NASA was having a honeymoon period with the Red Planet. An ambitious future was planned ...But then the Curse of Mars struck again

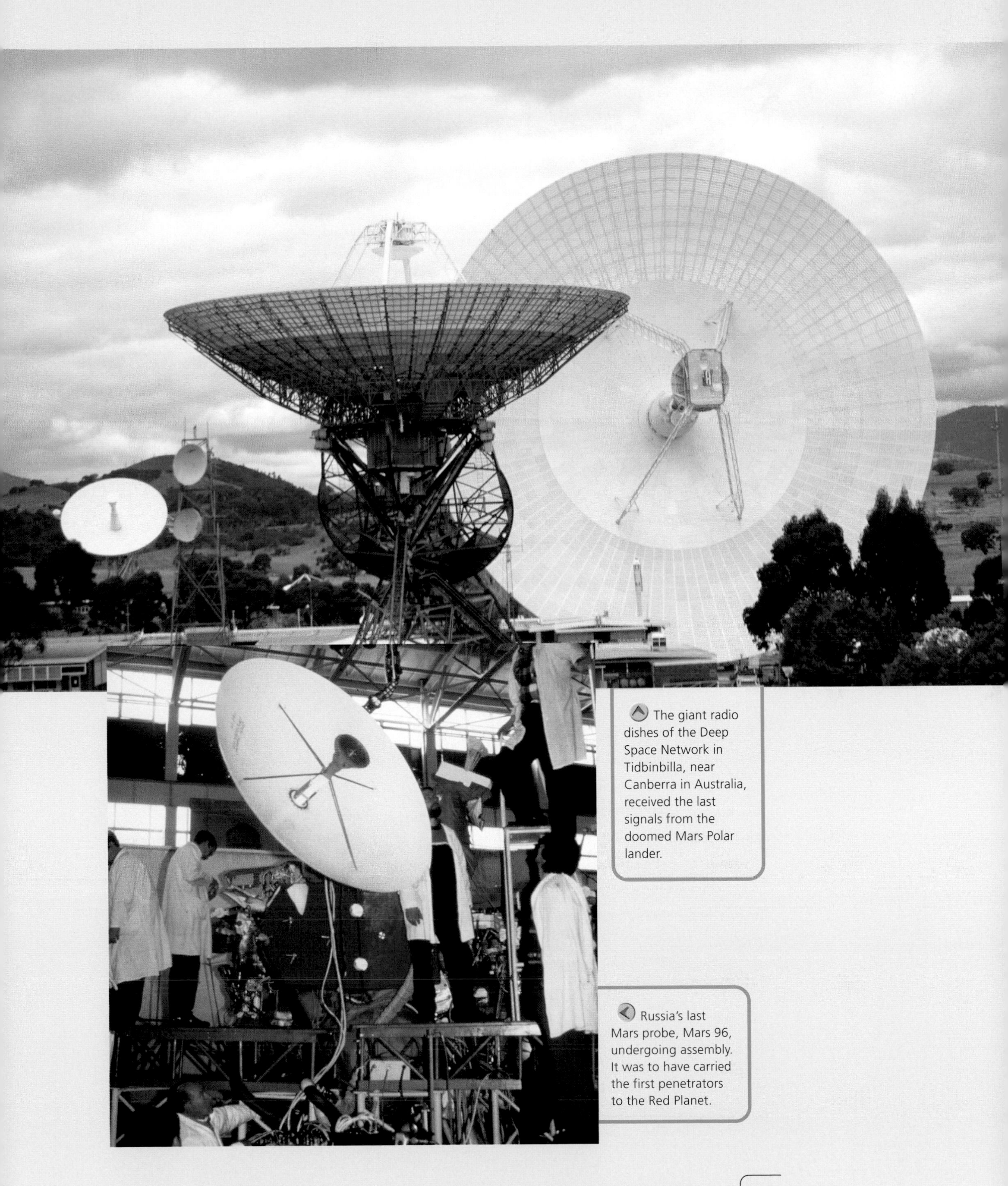

The giant radio dishes of the Deep Space Network in Tidbinbilla, near Canberra in Australia, received the last signals from the doomed Mars Polar lander.

Russia's last Mars probe, Mars 96, undergoing assembly. It was to have carried the first penetrators to the Red Planet.

layered terrain at the pole, you could have 100,000 years of geologic history there for the reading.'

On 23 September 1999, after a mad dash of only 286 days (shaving two months off the normal journey time), the Mars Climate Orbiter arrived at the Red Planet. Controllers at JPL fired an engine on the craft to push it into orbit. It disappeared behind Mars for twenty-five minutes, and, on emerging, should have sent a signal. The engineers heard nothing. 'The spacecraft was entering its corridor somewhat low,' commented the project manager ominously.

'You're always gonna get dumb mistakes, but what you have to do is have a system that catches them'

The giant dishes of the Deep Space Network were pressed into service to search for the missing craft. But after twenty-four hours – still nothing. The search was called off, and NASA had to admit to the loss of another Mars probe. Richard Zurek felt the loss very personally. 'These missions involve the work of literally thousands of people. And for a dedicated corps – those working on the spacecraft – this is what you've been doing for the last three or four years of your life. So there was a personal tragedy, but there wasn't really much time to focus on that because there was another spacecraft that was rapidly approaching Mars, the Polar Lander. And all our attention was focused on trying to make that work.'

And everything seemed to be going well for the Mars Polar Lander. The craft and its two innovative microprobe penetrators were in good health. We were over at JPL just a month before the scheduled landing, and even the mood of the place appeared to be picking up.

But on 3 December 1999 history cruelly repeated itself. The lander was due to have touched down about 500 miles from Mars's south pole and then signal to its controllers that it had landed safely. The signal never came.

This time, the controllers tried for days to contact the craft. 'Our confidence is less and less that we landed successfully,' confessed the project manager, after several attempts to coax a response out of the Polar Lander had failed.

It seemed incredible for a hi-tech agency like NASA to lose two state-of-the-art spaceprobes within a few months. Review boards were set up, procedures scrutinised, contractors thoroughly investigated, and weighty reports published. It all added up to a staggering tale of technical and managerial incompetence.

Veteran Mars geologist Mike Carr was on one of the review boards. 'I think we were pushing too hard, with too, too, too little resources. The whole system got stressed, people were working long hours, we didn't have adequate checking in place through inadequate funds. They tried this faster, better, cheaper approach. The cheaper part was pushed too hard. And the mistakes we made were dumb mistakes. You're always gonna get dumb mistakes, but what you have to do is have a system that catches them.'

The Mars Polar Lander succumbed to a combination of faulty sensors and software which together shut down the descent engines too early – while the craft was still 130 feet above the ground. 'It would then freefall to the surface,' observes Richard Zurek, 'perhaps even turning over, because without the rockets firing, it's not stable.' It seems that the sensors misread vibrations from deploying the three landing legs as the shock of touchdown, and the computer failed to spot the error.

The 'dumb mistake' in the case of the Mars Climate Orbiter was even more unfortunate.

Its manufacturers – Lockheed Martin – provided NASA with reference materials for the spacecraft's navigational systems in imperial, as opposed to metric, units (which NASA had requested). But no one bothered to check. Suspicions should have started to surface when the severely depleted team of navigators reported that the spacecraft was going increasingly off course.

NASA's John Rummel, Planetary Protection Officer, feels that controllers should have heeded the warning signs here. 'That was obviously someone not checking at a very critical point in software development, and navigators not getting their say when it came to modifying their spacecraft trajectory because they were uncertain. Navigators being uncertain is a bad thing. I know, I've been a navigator and it doesn't make them feel any better, either.'

In the event, the navigators were overruled, and the Climate Orbiter dipped too low into Mars's atmosphere in the attempt to insert it into orbit. Instead of skimming the atmosphere ninety miles up, it came in only thirty-five miles above ground. It almost certainly burned up.

All these findings are in the official report, but we uncovered an additional factor contributing to the demise of the Mars Climate Orbiter in a most unexpected way. We were scheduled to fly out of Denver to Washington in the summer of 2000. Our plane started to push back towards the runway, but the captain explained that the on-board radar wasn't functioning properly: we'd have to return to the stand. For three hours, we waited while the radar was put through its paces, but to no avail. It turned out that it had been struck by lightning on its previous landing.

We were all transferred to an already crowded plane – but fortuitously we found ourselves crammed in next to a young man who was formerly one of the senior engineers at Lockheed Martin working on the two doomed Mars missions.

He told us that he broadly agreed with the conclusions of the final report, but added some fascinating perspectives from the contractor's point of view. Most subcontractors in the US, he pointed out, still work in imperial units only – something that NASA, which uses metric units, should have checked. He criticised management and teamwork failures, and felt there were too few people involved – although the bright, dedicated core teams were working eighty-hour weeks. And apparently the penetrators for the Polar Lander – which Lockheed Martin was integrating with the probe – arrived late and hadn't been properly tested.

But the bombshell he dropped concerned the software. It was developed very late and it was undertested. We asked him if the failure of the computer on the Mars Polar Lander (which he described as 'just stupid') could be a casualty of the late software development. He didn't disagree.

NASA has now realised that 'cheaper' is one thing, but 'cheapest' is another. In the year 2000, it raised its Mars budget by $20 million to $250 million, and – at the same time – appointed a 'Mars Tsar' in charge of the whole Mars programme at its Headquarters. Scott Hubbard, praised for planning the enormously successful Mars Pathfinder mission, was keenly aware of his role. 'We're learning from experience. Part of my job now is to see that we stay in the centre of the mission highway, and that we don't fall off into the ditch on either side.'

Having spacecraft crash on Mars is bad enough. But what about Mars-bound spacecraft

Having spacecraft crash on Mars is bad enough. But what about Mars-bound spacecraft crashing on Earth – and, in doing so, threatening life on our planet?

crashing on Earth – and, in doing so, threatening life on our planet?

Mars 96 was Russia's last attempt to send a spacecraft to Mars. Designed in collaboration with the European Space Agency, it was an all-bells-and-whistles job, weighing seven tons, which actually comprised five separate probes: a huge orbiter, two landers and two penetrators. The craft was carrying seven ounces of plutonium in its batteries to provide power on the Martian surface. A hot radioactive element, plutonium is one of the most poisonous substances known.

Because the craft was so heavy, it needed more than even the mighty Proton three-stage rocket to loft it into space. The Russians added a fourth stage to kick Mars 96 in the direction of its target. The launch – which took place early on Saturday 16 November 1996 – was successful. All three sections of the Proton fired successfully, as did the first burn of the fourth stage. All it needed now was a second firing of the fourth stage to send it on its way to Mars. In the event, the fourth stage failed, and Mars 96 was trapped in an elongated Earth orbit with no chance of escape.

Mission controllers were horrified. The orbit indicated that Mars 96 – with its lethal cargo of plutonium – was destined to fall to Earth very soon. They desperately tried to bring it down in a controlled crash. Meanwhile, the American military were also aware of the situation. Over the weekend, both countries continued to track what they thought was the probe with the fourth stage still attached, and President Clinton personally phoned the Australian Prime Minister, John Howard, to warn him that it was headed for Australia – and that the US were more than happy to assist in any search and recovery operation.

It missed Australia, spared New Zealand, and finally burned up over the Pacific Ocean somewhere near Easter Island. Sighs of relief all round – except when the Russians called a press conference on the Monday. They revealed that the probe and the fourth stage had actually separated very soon after the launch. The object everyone was tracking, they now admitted, was almost certainly the fourth stage on its own. This was seen to have broken up over the ocean – but if that was the case, where was the plutonium-filled probe?

Jim Oberg – veteran watcher of Russian space antics – is in no doubt. 'Well, I kind of feel guilty that I'm the only person on Earth who had fun with Mars 96,' he tells us. 'And to reconstruct what happened to it took a lot of old skills and stuff that I thought were obsolete. After the worldwide panic over what turned out to be the booster rocket, not the probe, it began to fall into place that the probe had never been seen at all. It had fallen out of orbit only hours after launch, and because of my technical background in Mission Control in Houston, I was able to make some initial estimates of where it could fall without being seen by tracking devices.

'And later on, the final crucial confirmation came when people I knew in Chile collected and sent on to me eyewitness accounts of a fireball that had been seen crossing the Chilean coast at the exact moment that the probe was burning up. They were watching the probe, and it crossed the coast heading inland, towards the Andes Mountains.'

Oberg is certain that Mars 96 lies somewhere in the Chilean Andes. Fragments may have travelled as far as parts of Bolivia

The sky was completely clear over Chile that night, and thousands of people are likely to have seen Mars 96's demise. But the Chilean government refused to take reports seriously, having been assured by

Moscow and Washington that the probe burned up over the Pacific.

Oberg is certain that Mars 96 lies somewhere in the Chilean Andes. Fragments may have travelled as far as parts of Bolivia. But he suspects that much of the probe has survived, pointing out that the landers were designed to reach Mars's surface, and the penetrators even to withstand an impact with the ground.

'The question people ask me is, "Why hasn't it been found yet?" Who says it hasn't been found? That's the danger – it may have been found but not recognised. The plutonium is not going to wipe out a country or even a county, but the material in those batteries is hazardous to the person who opens them. The Russians say they've built this hard, so it could impact the ground and not break open. Well, yeah – it won't break open on impact with a rock, but it will break open on impact with a chisel.

'And the tradition in South America and in many Third World countries is to collect scrap and reprocess it. We've had a number of very, very nasty tragedies in which improperly disposed-of radioactive materials – health materials mostly – fell into the hands of children who unknowingly opened the material, played with it, and in many cases died from it.

'That material is on the ground, somewhere within long hiking distance of the Maoist guerrilla camps in Peru. How's that for a happy thought? What could they do with a couple of hundred grammes of plutonium?'

Despite calls by Oberg to the US government and nuclear safety officials to investigate his claims – maintaining that it is irresponsible not to go and look for Mars 96 – the official response is 'no comment'.

'Years after the crash, both the US and the Russian governments are very happy that the world would like to believe that the probe fell into the Pacific Ocean – which it didn't. There is no initiative to find out where it's got to. The initiative is not to find out.'

The Curse of Mars is alive and well – and it's on Earth.

The mysterious valleys of the Andes hide lost civilisations – and perhaps the missing plutonium-powered Mars 96 probe.

RETURN TO MARS

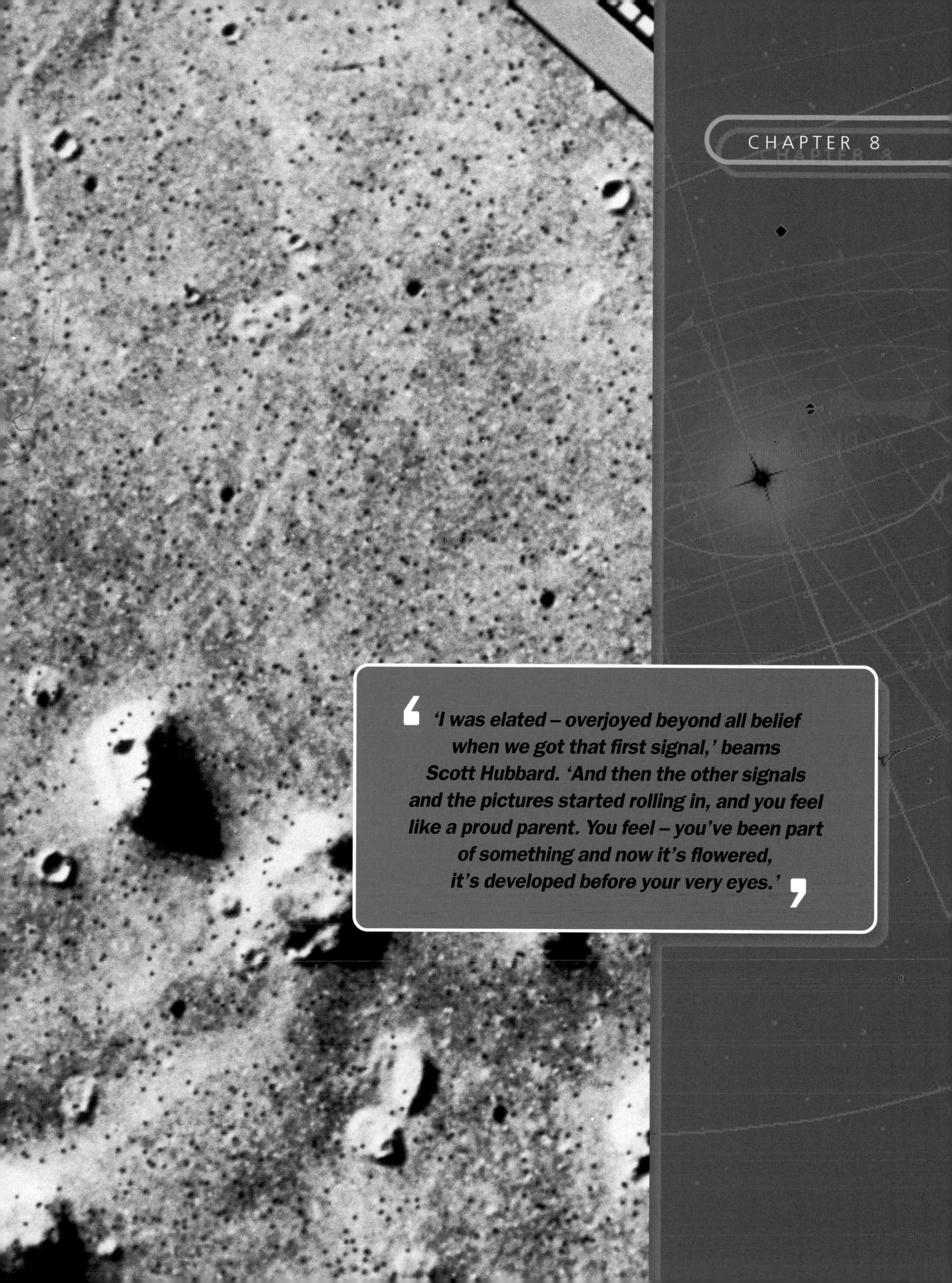

CHAPTER 8

'I was elated – overjoyed beyond all belief when we got that first signal,' beams Scott Hubbard. 'And then the other signals and the pictures started rolling in, and you feel like a proud parent. You feel – you've been part of something and now it's flowered, it's developed before your very eyes.'

Independence Day, 1997. While the rest of the United States holidayed, NASA engineers and scientists were distracted from celebrating 221 years of American home rule. Their hearts and minds were far away from their home country – 120 million miles away, to be precise.

Careering towards Mars was NASA's latest mission to the Red Planet. Pathfinder was perhaps oddly named. After all, previous NASA missions had flown past the planet, orbited Mars and even landed on its ochre plains. But after the success of the 1976 Viking missions, things had gone horribly wrong. Russian missions had failed; NASA's ambitious Mars Observer had mysteriously disappeared on arriving at the Red Planet. All eyes were now on Pathfinder, as the mission that might lift the twenty-one-year-long Curse of Mars.

Pathfinder came from a humbler background than Viking. No longer was NASA throwing billions of dollars at Mars. Manning's team had had to devise a simpler option – to fit NASA's new aim of 'faster, better, cheaper' missions.

They didn't have the luxury of parking in orbit around Mars, and then descending to the surface on heavy and expensive retro-rockets. Instead, it was a frontal attack on the planet. On 4 July 1997 Pathfinder hit the atmosphere of Mars at 16,600 miles per hour – eight times the speed of a rifle bullet.

A 360-degree Martian panorama, as seen by the Carl Sagan Memorial Station (the Mars Pathfinder lander). Parts of the spacecraft and its airbags are in the foreground; centre-stage is the Sojourner rover.

Rob Manning was the chief engineer. 'We hadn't done a mission to Mars in over twenty years,' he recalls. 'At the time of Viking, I was still in High School. NASA didn't have any experience remaining for such missions, and we were a rather naive group of young people pulling this together. Much to our surprise, we did figure out how to build and test a system that would actually land on Mars.'

Friction turned the spacecraft into a brilliant meteor lighting up the pre-dawn Martian skies. It was a nail-biting time for Scott Hubbard. Working at NASA's Ames Research Center, he had developed the heat shield that was intended to protect Pathfinder from the surrounding inferno.

'I had worked with the mission from the initial concept,' says Hubbard, 'and so – to be there on landing day! Everyone was so nervous – they were walking around on eggshells.'

Far away at Mars, Pathfinder had slowed its breakneck speed. The on-board computer shed the charred heat shield and deployed a vast billowing parachute. As Pathfinder swung down in the dark Martian skies, no one at Mission Control was any the wiser. The spacecraft wasn't programmed to report back until it had landed successfully.

Eight seconds from impact on Mars, Pathfinder suddenly blossomed. Huge white airbags – like the safety airbags in a car – swelled all around the triangular spacecraft. Resembling nothing so much as a vast bunch of grapes, Pathfinder hit the Martian surface – and bounced. It bounced, and bounced again – more than a dozen times in all – before settling on the red soil.

Pathfinder hit the Martian surface – and bounced. It bounced, and bounced again – more than a dozen times in all – before settling on the red soil

Euphoria erupted in the Control Room. Grown men and women hugged each other and wept. The Curse of Mars had been vanquished.

And the mission's success was summed up in a change of name. Now no longer a roving spacecraft but a fixture on the Martian soil, Pathfinder was renamed by NASA the Carl Sagan Memorial Station. The pioneering American astronomer and outstanding science populariser had died a few months earlier. NASA's Administrator explained the honour: 'Carl Sagan

was a unique individual who helped young and old alike to dream about the future and the possibilities it may hold.'

The bags deflated as the Sun rose on Pathfinder. Powered up by the Sun's rays, the spacecraft sent its first call home…

'I was elated – overjoyed beyond all belief when we got that first signal,' beams Scott Hubbard. 'And then the other signals and the pictures started rolling in, and you feel like a proud parent. You feel – you've been part of something and now it's flowered, it's developed before your very eyes.'

The cameras on the Sagan Memorial Station revealed a bleak terrain of rugged rocks. On the horizon lay a pair of hills, nicknamed Twin Peaks. So far, the mission had done no more than the Vikings had achieved. But all this was to change. On board the spacecraft was a six-wheeled rover, the first Martian vehicle to venture away from its landing site.

When it came to selecting a name for the rover, NASA had turned to the nation's schoolchildren. The shortlist included the Greek goddess Athena, the Native American Sacajawea, who guided the explorers Lewis and Clark, radioactivity pioneer Marie Curie, and Judy Resnik, an astronaut killed in the 1986 explosion of the Challenger space shuttle.

The microwave-oven-sized Sojourner rover nuzzles up to a Martian rock to test what it's made of. The solar panel on its back provides its power.

But the winner, twelve-year-old Valerie Ambroise, selected a lesser-known heroine. Sojourner Truth – originally Isabella Van Wagener – was an African-American campaigner during the Civil War. Ahead of her time, she made it her mission to travel up and down the land to advocate the rights of all people, black and white, male and female. And so the rover, travelling up and down a stranger land, became Sojourner.

Rob Manning was one of the team driving Sojourner. 'It was surreal, and sublime,' he recalls, 'with this very eerie orange glow and the strange surface textures. It's almost as if you were there. Right now, Mars is feeling a long way off, but in July and August of 1997 Mars was right at our fingertips.'

Manning's team drove the microwave-oven-sized rover round the nearby rocks, and nuzzled Sojourner up to them to test what they were made of. The cuddly robot – and the successful return to Mars – made an instant hit of the mission – aided by the timely arrival of the Internet. All over the globe, people logged on to the Pathfinder website to experience the Red Planet at first hand.

'That's another thing that surprised me,' confesses Manning. 'We were enthusiastic about landing on Mars, but it's another thing altogether when you realise there are many people around the world who find this really interesting. In six months, I think our website received almost a billion hits.'

For geologists, the biggest surprises were the boulders that Sojourner investigated. They'd expected Martian rocks to be dark basalts, solidified lava from the vast low-profile volcanoes that resemble the gently oozing Hawaiian volcanoes on Earth. Instead, Sojourner found they were lighter in colour: chemically, they resembled rocks thrown out by explosive volcanoes.

The mission was intended to last a month, but Sojourner was still going strong three months later, when the Sagan Memorial Station ran out of power. Sojourner's only link with Earth was through the lander; in the event of a communications failure, the rover had a built-in command to keep circling it until the link was restored. The final act of the Pathfinder mission must have been an orphaned Sojourner endlessly trundling round and round the defunct lander...

By that time, another robot craft had arrived to help prise open the secrets of Mars. High overhead, the Mars Global Surveyor was orbiting

The final act of the Pathfinder mission must have been an orphaned Sojourner endlessly trundling round and round the defunct lander

the Red Planet. The Global Surveyor was a reincarnation of a previous NASA Mars probe, the ill-fated Mars Observer that had failed as it approached Mars in 1993.

'I've been involved in this mission basically for seventeen years,' says Arden Albee as we catch up with him in 2000. 'I was involved in the early studies that became Mars Observer. When we lost that mission, we developed Mars Global Surveyor as a faster, cheaper version. It particularly picks up on the idea that you put it into a circular orbit, let Mars spin under you, so you systematically map the entire planet.'

But as Global Surveyor approached Mars in September 1997, the Curse of Mars loomed ominously. To slow the craft down, it would skim the top of Mars's thin atmosphere. This aerobraking manoeuvre would be helped by the drag from the big solar panels sticking out from it.

Like huge wings, the solar panels had hinged open after the spacecraft was launched from Earth. 'But a funny thing happened on the way to Mars,' Albee recalls. 'When we started aerobraking, the wings began to flutter.' One of the solar panels had not latched into the open position; now the rush of air past it was threatening to push it shut. Mission planners devised an emergency rescue strategy: turn the spacecraft round, so that the aerobraking would tend to push the panel the other way, into its 'open' position. But that meant a wait of half a Martian year; almost a year on Earth, before Global Surveyor reached its final orbit.

'We used to call it the "stealth mission" to Mars,' jokes Bill Hartmann. 'It got there the same summer as Mars Pathfinder, and everybody paid attention to that and Sojourner – which was great – and we got there two months later and nobody was interested. But it's a bit ironic, because Global Surveyor has now been such a great success.'

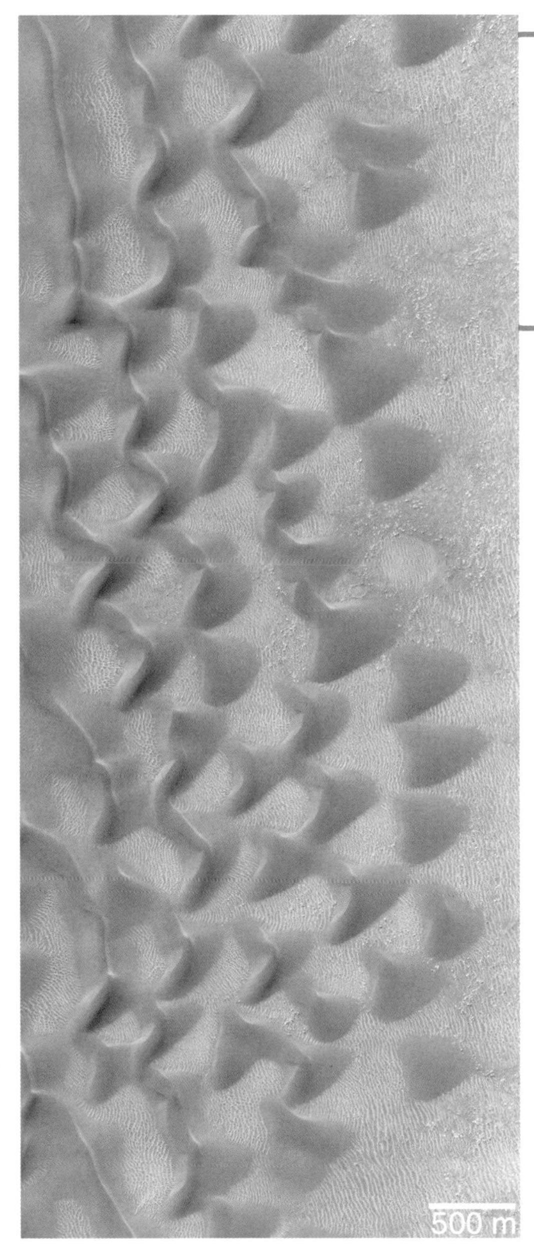

The 'Sharks' Teeth' sand dunes in the Martian desert, imaged by the Mars Global Surveyor. The scientist who built the spacecraft's camera is stirred to call this view 'just plain pretty'.

Over the years it's been in orbit, the Mars Global Surveyor has certainly told us more about Mars than any mission since the Vikings. Albee runs through some of the new discoveries for us.

'Beforehand, we didn't know if Mars had a magnetic field,' he begins. 'Now we've found that Mars doesn't have a global magnetic field, but it does have small patches of strong magnetism that date from the very early stages of Martian history.'

'The formation of the company was not really designed for furthering private exploration of space,' he reveals, 'so much as for furthering Mike Malin's exploration of space!'

The Global Surveyor has also measured the shape of Mars with uncanny precision, mapping its contours in intricate detail. 'We now have the best topography of any planet,' Albee boasts, 'including the Earth!' To get from the bottom of the huge Hellas basin, blasted out by an ancient impact, to the top of the giant volcano Olympus Mons you'd have to climb twenty miles upwards.

If the Global Surveyor mission has done one thing, it's to make Mars seem even more of an alien world – a planet we can't understand just as a brother of the Earth. And it's the pictures from the Global Surveyor's orbiting cameras that have done most to turn our ideas of the Red Planet on their head.

The cameras on the Global Surveyor are the brainchild of Mike Malin. A Mars enthusiast since childhood, Malin went to work for NASA after university. 'As an undergraduate, I was in physics, and then I watched the television coverage of the Mariner 6 and 7 fly-bys of Mars and was pretty much hooked by what I saw.'

Studying Mars meant turning from physics to geology, and Malin saw a problem that was bedevilling the geologists' attempts to understand the Red Planet. Whether they were looking at pictures from the early Mariners or the Vikings, geologists were seeing either a very broad-brush view from orbit or a detailed picture on the ground. What was needed, he decided, was a telephoto reconnaissance of Mars from space.

'I was a member of the science working group on a mission that became the Mars Observer,' Malin explains, 'and with Ed Danielson at Caltech I proposed a camera for that mission.' He attributes the success of the Mars Observer Camera not just to its design, but to his decision to work outside NASA's bureaucracy. 'In a large institution there's this demand that you bow to its will. I'm not the kind of person who bows to anybody's will.'

So he set up his own company, Malin Space Science Systems, based at San Diego in southern California. 'The formation of the company was not really designed for furthering private exploration of space,' he reveals, 'so much as for furthering Mike Malin's exploration of space!

'But we do have a more altruistic view as well,' Malin smiles, 'which I hope we have transferred to my employees – that you do better by having scientists intimately involved in the development of instrumentation and in the execution of a mission.'

The Mars Observer Camera flew to Mars in 1993 – and disappeared along with the rest of the spacecraft. When NASA rebuilt a stripped-down version of this mission as the Mars Global Surveyor, the Mars Observer Camera was once again aboard.

'What's the most amazing thing to me,' Malin says, 'is that the camera worked. To translate a scientist's concept of a tool he wants into this thing at Mars that we can use every day. And, you know, we can tell what something three to four metres across is.' The camera could distinguish between an estate car and a four-wheel drive on the surface of Mars.

The Mars Observer Camera has shown sights that no one could have predicted. Even the conservative geologist Mike Carr is moved to remark: 'The new high-resolution imaging from Mars Global Surveyor is just very, very difficult to understand on the basis of the Viking imagery.'

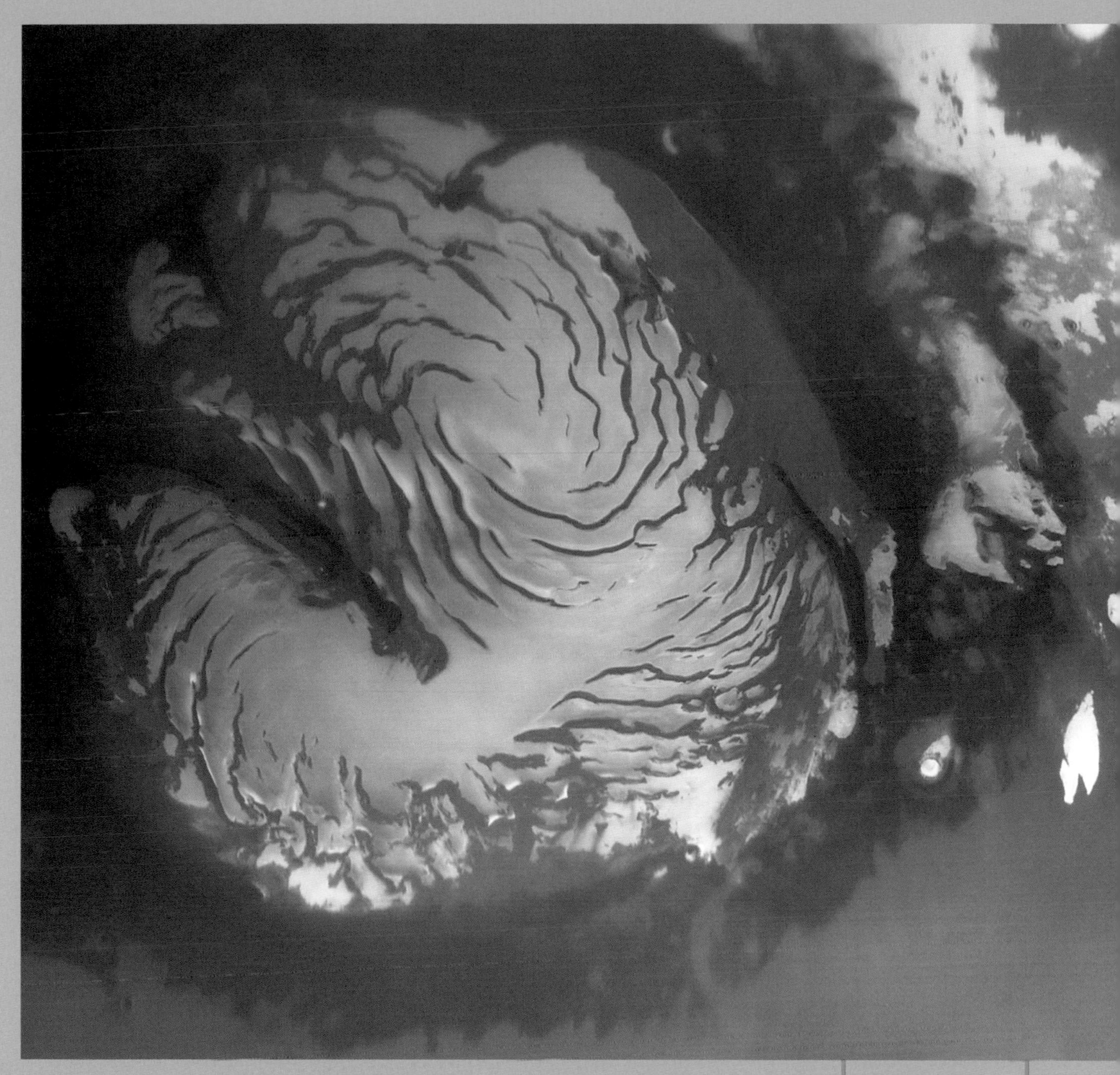

Like a giant ice-cream swirl, this is the icy mantle that adorns Mars's north pole during summertime on the Red Planet. The snow and ice stretch over 700 miles. The dark rim consists of sand dunes perpetually reshaped by the wind.

'Take a look at the polar caps,' Arden Albee expostulates. 'The north polar cap has layers – it's quite simple. But the south pole is crazy! It's got Swiss cheese whorls, and no one has any good explanation for them.'

And Malin's camera is responsible for most of the scientific exposés in this book, from the real visage of the Face on Mars, through the evidence for active volcanoes to the gullies indicating that streams of liquid water can still flow on this frozen world. And it has plenty of work still to do. Its focused stare on the planet has so far covered only 0.1 per cent of Mars's surface.

The Mars Global Surveyor will keep up its scrutiny of the Red Planet until it is able to hand

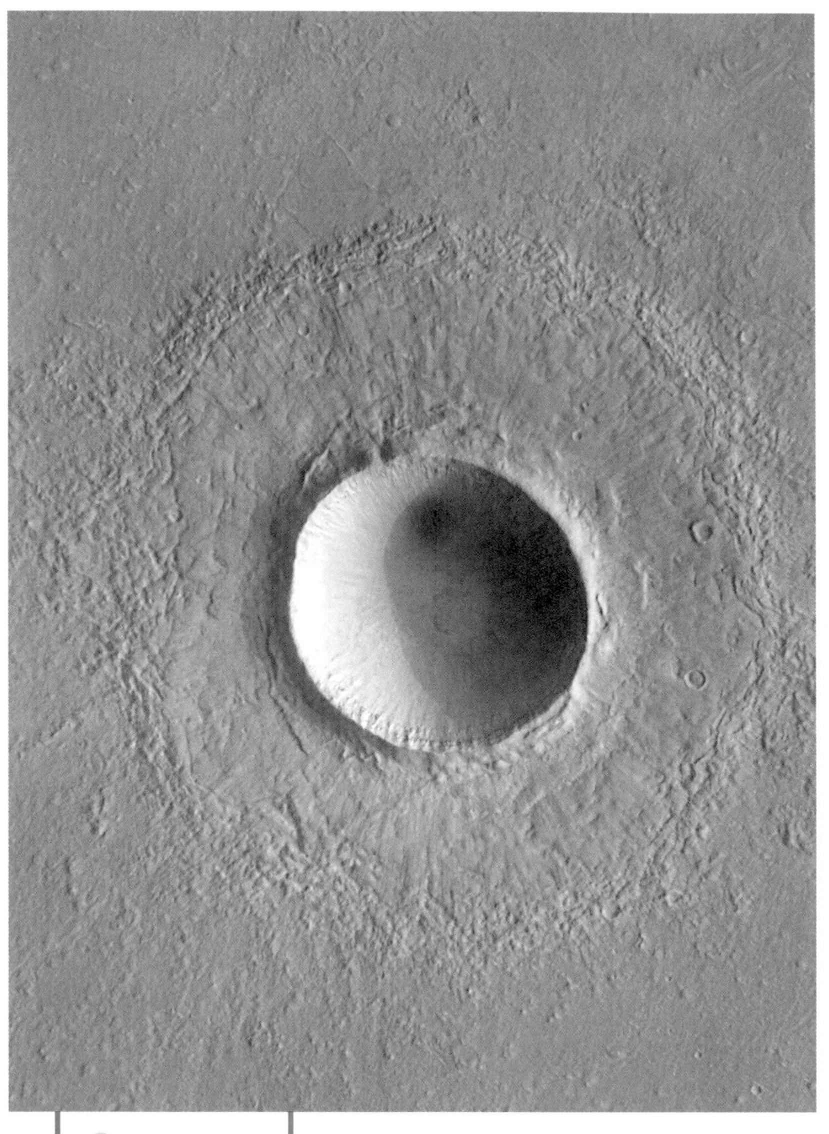

Small but perfectly formed – this crater is tiny by Martian standards, just a mile across. Even so, it is twice the size of the Earth's famous cosmic blast-out, Meteor Crater in Arizona.

on the baton to its successor, 2001 Mars Odyssey. Launched in April 2001, this orbiting craft again bears the legacy of the ill-fated Mars Observer that disappeared in 1993. This time it's an instrument to seek out whether Mars really does hoard ice – or even liquid water – in its surface. And there's a nod to the future. The spacecraft will measure the radiation levels near the planet, specifically to check how dangerous they will be to a future generation of visiting astronauts.

'The year 2001 has a special significance to many of us who recall the thrill of reading the book and watching the movie *2001 – A Space Odyssey*,' says Scott Hubbard. 'It seems fitting to name the mission 2001 Mars Odyssey not only in honour of the story and the movie but also to herald the start of a new long-term journey to explore Mars.'

After his early track record on Mars Pathfinder, Hubbard rose to great heights. 'I was the Mars Program Director at NASA Headquarters – and I'm also known as the Mars Tsar,' he chuckles.

Scott Hubbard is a counterblast to the stuffiness of NASA bureaucracy. After finishing university, he spent a year as lead guitarist in a rock band in legendary Nashville, Tennessee. When we meet him in NASA's Washington headquarters, the music playing in the background is Bach's Brandenburg Concertos – but Hubbard still cuts a flamboyant figure.

'As a young boy I looked through a telescope in the back yard in Kentucky where I grew up,' he enthuses,' and one of the first things I looked at was Mars – this tiny red dot. And so I've had a special interest in that planet for a long time.'

And as Mars Tsar he had a chance to do something about it. Hubbard was in charge of coordinating all NASA's unmanned expeditions to the Red Planet. The post was created after the humiliating debacle in 1999 when NASA lost two consecutive missions due to administrative foul-ups.

While we're there – in the summer of 2000 – Hubbard is about to announce what missions

'The year 2001 has a special significance to many of us who recall the thrill of reading the book and watching the movie *2001 – A Space Odyssey*...It seems fitting to name the mission 2001 Mars Odyssey'

In 2004 a freezer-sized rover will take its first spin on the Red Planet. One of two Mars Exploration Rovers, it will build on the heritage of the tiny 1997 Sojourner; but in one day this rover will travel as far as Sojourner travelled in its entire lifetime.

will follow the 2001 Mars Odyssey orbiter. And he's putting the concluding touches to a 'road plan' for the next ten years. People bustle in and out of his office; we hear the word 'budget' being whispered. As we've travelled the States, Mars researchers have only been able to express hopes for the future exploration of the planet. Now we are at the nerve centre, and at the moment those hopes are about to be dashed – or realised.

Hubbard has a diplomatic answer to our probing questions on the future of Mars exploration. 'We are in the process of reformulating the science, with "follow the water" as a strategy,' he replies. 'I think you'll see that we have a programme of robotic exploration that is more resilient. It may be fewer missions than in the past. And I think you'll see a programme that collects information and makes observations that are valuable – and even critical – to the eventual human exploration of Mars.'

Waiting on Hubbard's decision with more anxiety than most is Steve Squyres, of Cornell University at Ithaca in upstate New York – the university where Carl Sagan had worked. In more relaxed circumstances, later in the year, Squyres told us why.

'That whole thing was wild, just bizarre,' Squyres says. 'We had four instruments built – but no ride to Mars and no idea what was going to happen at all. We had this beautiful rover design, but the lander that was supposed to carry it was too closely related to the Mars Polar Lander and nobody wanted to fly it.'

The Mars Polar Lander was one of the two probes hit by the Curse of Mars in 1999. After these failures, all bets were off for missions beyond the 2001 Mars Odyssey. The chances to

International cooperation takes another step into interplanetary space, as NASA engineers test-drive a Russian-built 'Marsokhod' rover. Unfortunately, this tough rover is too heavy to send to Mars on any current rocket.

fly to Mars come at roughly two-yearly intervals, and NASA's plans for launches in 2003 and 2005 were totally thrown into disarray.

'Then some folks at JPL crocked up the idea of delivering a revised version of the rover,' Squyres explains, 'using the Pathfinder airbag system. We had just two weeks to study this idea and to convince NASA it was even credible.'

At a meeting in May 2000, NASA had over a hundred options to consider, including the new rover that had been added at the last moment. They shortlisted just three – including the rover. 'I was absolutely stunned,' Squyres recalls. 'Then NASA knocked it down to just two options – the rover or an orbiter mission – and it was head-to-head competition.'

Over at JPL, two rival teams of engineers got their heads down again. On one side, Rob Manning's team – working with Squyres as chief scientist – drew up blueprints for a bigger and better Sojourner. Mars meteorologist Richard Zurek headed a team that planned a replacement for the other 1999 failure, the Mars Climate Orbiter.

'Then in the middle of July, it all culminated in this one-day shoot-out at NASA headquarters.

I gave the science pitch for the rover; and Rich Zurek gave the pitch for the orbiter. And then they threw us out of the room and they voted...'

Squyres was stunned once more: Scott Hubbard and the top brass at NASA picked the rover. And a bigger shock was in store. The prospect so enthralled NASA's Administrator, Dan Goldin, that he revised the Agency's overall budget to make money available for a second 2003 rover.

'They took us completely off-guard,' says Squyres. 'We had to scurry around at a furious rate to come up with a cost estimate for them – the two-rover concept just blew me away. The cost difference isn't as big as we'd imagined, though.' Squyres explains that, to send one rover to Mars, the JPL team must build five altogether – the others being test models before the real thing flies. So for NASA to dispatch twin rovers to Mars, the total doesn't double but only rises from five to six.

To check out the latest model of Mars-mobiles, it's time for another trip to JPL. In a large echoing engineering lab, we find Eric Baumgartner. He's flanked by a pair of vast yellow pneumatic tyres, almost as tall as Baumgartner himself. Slung between them is the small metal box that forms the heart of a future Mars rover.

'This big inflatable rover is a new concept vehicle that's really meant to go over very diverse terrain,' Baumgartner explains. 'The nice thing about it is that you can suck it down into a very nice compact form to launch and cruise to Mars, then you can deploy it out.'

Baumgartner also enthuses about a design that the Russians have been testing. 'The Marsokhod rover has big conical wheels and is articulated in the middle, so it can pretty much roll over just about anything. But it is very massive – it would never fit on a launch platform that we could send to Mars now.'

Technicians at the back of the lab are testing out different kinds of wheeled robots on Mars. Well, not actually 'on Mars', but the next best thing

Instead, each of the 2003 rovers will fit inside a casing about the size of the Pathfinder mission, and bounce down to Mars in a bunch of airbags. But this time there will be no fixed lander. The rover itself is the only payload being flown to Mars; once it trundles out on to the red desert, it will be on its own.

As we speak, technicians at the back of the lab are testing out different kinds of wheeled robots on Mars. Well, not actually 'on Mars', but the next best thing. A huge sandy arena is scattered with Mars-sized rocks, against huge murals of the Viking view of Mars's surface.

Baumgartner's background fits him for research into wheeled vehicles. As a graduate student in engineering, he worked on robots for use on factory floors and developed a robot-assisted wheelchair. And he's keen to build on JPL's past success with Mars rovers. 'Sojourner couldn't go up steep slopes or get over big rocks,' he explains. 'So we're looking into rovers with more mobility.'

Rob Manning – the engineer behind the Sojourner rover – is also on the team designing the 2003 rovers. 'This one is truly a geologist that will go out and explore,' he explains. 'With the Pathfinder mission, the rover couldn't go very far. Now we'll be able to wander over the surface of Mars for some distance to explore different kinds of features, and we'll look closely at surface rocks and soils.'

'There's a small microscopic imager,' adds his colleague Mark Alder, 'a sort of hand-lens like a geologist might use to look at grains in the rock.

And there's a little grinding wheel that'll grind off a piece of the rock so we can look at the rock interior.'

'There's a trade-off,' says engineer Baumgartner, 'between the science and the engineering. They tell us what are the scientifically interesting sites, but we may have to say, "Sorry, you can't go there" or "You can only go so fast". I've certainly learned a lot about geology by being around these folks, and hopefully they've picked up a little bit of engineering as well!'

To help it find its way around, the new rover – the size of a large freezer – has intelligence built in. Two cameras on a mast at human eye height give it stereo vision, and its on-board computer builds up 3-D maps of the surrounding terrain.

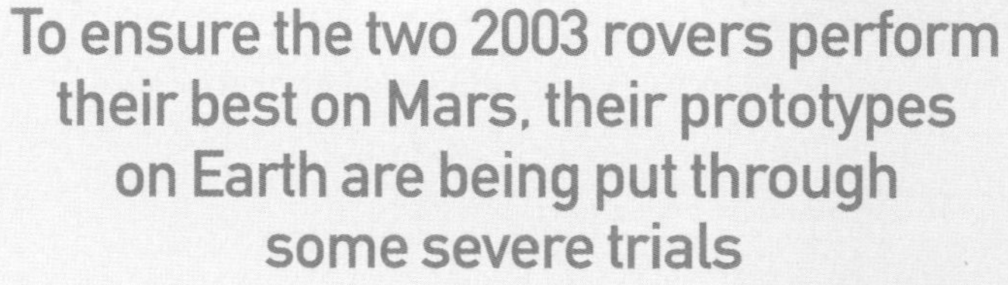

To ensure the two 2003 rovers perform their best on Mars, their prototypes on Earth are being put through some severe trials

'What we do,' continues Alder, 'is to give it commands at the beginning of the day. The vehicle then has its own software that's smart enough for it to look at the terrain and figure out how to get from here to there in detail. As it goes forward, it should be able to say: "Well, here's an obstacle; I'm going to go around it." The rover can do its own stuff for a day, but from day to day we command it from the Earth.'

To ensure the two 2003 rovers perform their best on Mars, their prototypes on Earth are being put through some severe trials – not just at JPL, but out in the desert where the engineers can simulate Martian terrains. Baumgartner has just returned from a relatively accessible Mars-on-Earth setting with FIDO – the Field Integrated Design and Operations rover. 'We took FIDO to the central Nevada desert, the great basin-and-range area, for a big field test. The rover was

Bringing it on home: an early NASA artwork for the Mars Sample Return Mission. Guided by mission controllers on Earth, the rover will dig out samples and place them in the six-compartment container. The rocket in the background will then propel the sample container back to Earth.

operating on its own out there, with a small field team, while the majority of the scientists and engineers were back here at JPL operating the vehicle as if it was on Mars.

> **NASA is planning the most audacious of unmanned Mars missions: the Mars Sample Return**

'In the morning we talked to the field team from JPL,' continues Baumgartner, 'and showed them the sequence we plan to send. They would review it and give us a Go/No-Go. If we got the Go, then we up-linked the sequence through a satellite dish and the rover'd start executing the commands. And then the telemetry, engineering data and images from the rover got fed back through the satellite link to JPL, as if the data were coming back through the Deep Space Network from Mars to Earth.'

The Go/No-Go from the field team is a safety net that won't be available on a real Mars mission. But Baumgartner is at pains to point out that most of the No-Goes occurred because of conditions that wouldn't happen on Mars. 'They're mostly weather-related: rain, high winds – and they had snow. But there was one instance where the field team said, "No, I'm sorry, we're not going to send that sequence through, because you're in a situation you don't understand right now".'

The JPL team shut down for the day and pored over the engineering data and images from FIDO. Eventually, they figured out what the problem might be. Next day, they contacted the field team with their conclusion. 'They said, "Yes, you're right",' Baumgartner smiles. '"The rover's hung up on a rock!"'

FIDO wasn't testing just the engineers. For a couple of days NASA invited schoolchildren in the US and Denmark to take charge of the rover over the Internet. They managed not to get stuck on a rock. The initiative marks the beginning of a bigger public involvement with Mars. 'We are teaching students that the future Mars explorer will be a tele-operator,' says Lou Friedman of the Planetary Society, a public outreach group, 'and for the first time we're going to be able to give students the chance to control a vehicle on the surface of Mars as part of the Mars 2003 mission.'

Today's scientists, as well as the engineers, were under the spotlight when it came to the FIDO tests. They were faced with describing the site entirely as experienced by the rover. Senior researcher Steve Squyres recalls the experience. 'We were in a windowless room in JPL and operated the rover blind, as if it were on Mars. In fact, it was in Nevada somewhere – but I don't know where!

'I'm serious,' he adds. 'My deputy is the one who picked the field site, and he's not going to tell us where the site was until we'd finished writing the paper that describes the result.'

Chris Chyba, of the SETI Institute in Silicon Valley, recalls a previous test that put the scientists on their mettle. 'There's a famous story about a rover test in the American south-west,' he says, 'and about how it drove right over the top of a fossil footprint – and nobody watching the screens back at JPL noticed it!'

NASA's next generation of rovers isn't intended to go fossicking, but Chyba's comment highlights the next phase of Mars exploration – the intense search for life, past or present. 'I think it's going to be very difficult to find fossils on Mars before people get there,' Chyba concludes. 'The irony is that, with robots, it may even be easier to find evidence for current life on Mars.'

With this thought in mind, NASA is planning

the most audacious of unmanned Mars missions: the Mars Sample Return. 'It's a tall order – to bring samples back to Earth reliably is a hard thing to do,' admits Rob Manning, who was chief engineer for the programme until it was postponed following the 1999 spaceprobe failures. It's slipped down NASA's programme from 2007 to 2014. But Manning is convinced it will happen: 'I think we should do it; we will do it; we have to do it to learn more about Mars.'

'The Sample Return mission is one of the most technically difficult things we're going to attempt in the robotic space exploration programme,' says his colleague Mark Alder as he talks us through the mission. In short, it involves a very large lander with a rover that will pick up several samples of Martian rock. The rover loads them into a canister aboard a rocket, which shoots it up to Mars-orbit. An orbiting rocket picks up the canister and propels it back to Earth, where it's put into a special re-entry vehicle that will land it safely in the desert.

'This is a very long, perhaps three-year strategy,' says Alder, 'not unlike the Apollo Moon missions, where we had a lander going down to the surface of the Moon, coming back up and meeting up with an orbiter and the orbiter bringing the astronauts back to Earth.'

Apart from being a robot mission, there's one big difference from Apollo. The astronauts were returning from a sterile Moon, but the Mars spacecraft might – just possibly – be carrying bugs from the Red Planet. And who's to say that these microbes are safe? Standing between us and a possible invasion of Mars bacteria is NASA's Planetary Protection Officer.

The person who bears that title is John Rummel. For five years an officer in the navy, Rummel started out his scientific life with an interest in biology and ecology – how different living systems depend on each other. 'When I was a freshman in High School,' he recalls, 'I was interested in making a life-support system, and I did some experimentation that involved a bunch of algae and two unfortunate frogs...'

Rummel has moved on from responsibility for a pair of frogs to responsibility for a pair of planets. We catch up with him in the labyrinth of NASA's Headquarters, where he works closely with 'Mars Tsar' Scott Hubbard in safe-guarding the future for the worlds of our Solar System.

Rummel reassures us on what will happen to that sample of Mars when it lands on our planet. 'The plan now is to bring it back to Utah, of all places,' says Rummel, 'and put it into a containment facility. Here it would be assayed in a couple of ways, doing some basic tests to find out whether or not there's any indication of a biohazard of any kind.'

The containment facility will be as secure as the labs where scientists study lethal and contagious viruses, like Ebola. The quarantined sample will be tested with a battery of biological procedures. Scientists will attempt to culture any microbes in standard lab dishes, and try to infect eggs. With an eye to infections that proliferate in unusual ways, NASA's French colleagues have also insisted that the team includes experts on BSE.

Rummel has moved on from responsibility for a pair of frogs to responsibility for a pair of planets

'I think that life exists in the Universe. It certainly has the potential to exist on Mars, so I wouldn't be surprised to find life somewhere on Mars,' says Rummel. But – to our surprise – he's fairly laid back about the idea pathogens – at least obligate pathogens – don't exist there.'

Obligate pathogens are germs that have evolved along with humans – and account for almost all human diseases. And other kinds of microbe are not likely to be deadly: 'If there are adventitious pathogens – something that could inconsequentially infect humans – then we'll be looking out for those. It could be that it would just co-exist with all the other Earth microbes and not do any harm.'

In fact, Rummel's greatest fear has nothing to do with genuinely alien microbes – but with Earthly bugs transported to Mars on previous spacecraft, and mutated in Mars's harsh environment. 'If Earth-life was taken there years earlier, and then we bring it back, I'd want to check it to make sure it was OK. We know there are lots of things on Earth that are incompatible with human health if they're introduced in the wrong place. In history we can see biological problems caused by biological invasions – I mean, for example, thanks to the devil of the sailing ship.'

Nicknamed the 'stealth mission' because of its low public profile, Mars Global Surveyor has stunned scientists with its detailed views of the Red Planet. Single-handedly, it has turned the science of Mars on its head.

'My role on these missions is always Dr No,' Rummel jokes. 'I mean, I tell people what they can do and what they can't do'

that the returned Mars-rocks might trigger a 'Mars-plague'.

'It's extremely unlikely,' he continues, 'that life on Mars would have any properties that would be incompatible with the health of the Earth's biosphere or with human health. There are no humans on Mars so human

That's one reason why Rummel sees planetary protection not just as an issue for the future but as a day-to-day responsibility. 'Right now we're sending missions out to a variety of different worlds that could support Earth-life,' he says, 'and I want to make sure that when we send them out they're clean enough that Earth-life won't contaminate other planets.

'My role on these missions is always Dr No,' Rummel jokes. 'I mean, I tell people what they can do and what they can't do. And within certain confines I have full sway.'

A spacecraft may look gleamingly clean when it's built, but Rummel reveals it may harbour as many as 100 million microbes. 'When we put together a spacecraft to go to Mars, we clean it assiduously and reduce that bio-load down to roughly 300,000 microbes – or even down to 30,000. It's less biological contamination than you'd get in a portion of caviar.'

For the 1976 Viking missions – built specially to search for Martian life – NASA was even more concerned that any life they detected wouldn't just be hitchhiking bugs from Earth. So they reduced the potential cargo down to a mere thirty or so microbes by baking the entire spacecraft before launch: 'Viking was your basic spacecraft casserole,' says Rummel. 'You encapsulate it first, bake it later and then the surrounding capsule keeps it clean.'

Viking found that Mars has a natural sterilising environment. Its soil is very dry and contains natural chemical bleaches, while the surface is bathed in ultraviolet rays from the Sun. As a result, later missions like Pathfinder and the Sojourner rover have been cleaned only to Rummel's 'caviar level'.

But the discovery of warm, wet regions – cosy niches for life, whether indigenous or imported – is putting Rummel back into Dr No mode. 'If we go to one of these watery places,' he cautions, 'things would have to be as clean as Viking.'

Despite such cautions, NASA sees the Sample Return project as the most important mission before humans travel to Mars themselves. Everett Gibson from the Johnson Space Center in Houston is desperate to get a sample in his lab. 'We could collect a variety of pebbles or small samples the size of a marble, so we're bringing back – say – twenty to fifty different rock types. We may get a sample of a carbonate, a limestone, a granite – or a sample with a fossil.'

Gibson was one of the NASA team that announced 'fossils from Mars' in 1996. The discovery is still controversial – if only because the supposed fossil was in a volcanic rock, not the kind of place you'd look for fossils on Earth.

'The key question is to get a piece of sedimentary rock,' says Chris McKay – 'Mr Mars' to his colleagues at NASA's Ames Research Center – 'because sedimentary rocks – for example, the bottom of an ancient lake – are much more likely to preserve good, clear fossil evidence of life.'

Pillinger is convinced that his British mission can solve the question of life on Mars well before NASA even gets its massive and expensive Sample Return mission under way

And Bill Schopf – discoverer of the oldest fossilised life on Earth – sees sample return as essential to understanding if life ever existed on the Red Planet. He doesn't think that robots on Mars will ever answer the question. 'Let's imagine there were giraffes or elephants running around on the Martian surface. Then a robotic mission could take a picture of them, show them running and hear them trumpeting – and you'd say, "Oh my golly. There's life on Mars – that's terrific!"'

But there aren't elephants or giraffes on Mars, and the best that Schopf can hope for are microscopic organisms, or the rocky deposits they might make. He contends that photographs will never reveal whether deposits on Mars are made by living cells or just by geological processes. 'On the other hand, when we bring a bunch of rock samples back and study them in the laboratory, then I think we can solve it.'

But not everyone is swept along by sample-return mania. Veteran Mars geologist Mike Carr advises caution. 'I'm a strong advocate of sample return, but there are two camps. One is that we should get the samples as soon as we can. But there's another camp that says we're going to get so few sample return missions that we should make a very intensive exploration of the planet before we commit to going to a specific site. This battle is going on now – and it's very passionate.'

'To carry out the sample return that people have envisioned requires technology we don't have today,' adds fellow Martian geologist Bruce Jakosky. 'We're still developing the ideas, let alone the hardware.'

'Sample return is not easy,' continues Carr. 'It will consume all the resources of our Mars programme – and it's risky. And it's not clear that we really know how to do it. Some people think that the current concepts have a high likelihood of failure.'

'If you pump more money into it,' says Jakosky, 'you can do it quicker. At the current level of resources it could take two decades. If we have to send a dozen missions to be sure of finding a place where life might have lived, then maybe it's three decades. And three

On the trail of Martian life, Britain's Beagle 2 is put through its paces on Earth. Unfolding from a circular shell, Beagle deploys its blue solar panels, and stretches out its PAW (right) to test the Martian air and rocks.

decades is no different in my mind from an infinite time.'

So do we have to return those samples at all? Colin Pillinger thinks not. A leading planetary scientist, Pillinger is convinced that his British mission can solve the question of life on Mars well before NASA even gets its massive and expensive Sample Return mission under way. The innovative spacecraft is called Beagle 2.

'It's named in honour of the original *Beagle*,' Pillinger explains, 'which was the ship that took Charles Darwin around the world and led to our understanding of evolution.' Again following the Darwin trail, Pillinger is proud that Beagle 2 is a British enterprise. 'We see the idea of Britain going to Mars as being so important, and the idea of looking for life on Mars is a question that unites everyone.'

We find Pillinger on the campus of Britain's Open University in Milton Keynes. The Open University is a premier distance-learning establishment, and – though it doesn't have a bevy of busy undergraduates on site – the university does need a campus for its researchers.

In his cramped office, behind a huge inflatable hanging globe of Mars, Pillinger takes pains to point out that Beagle 2 will land on Mars well before NASA's enormously complex Sample Return mission. In fact, the British mission will touch down even before NASA's next landing mission with its pair of rovers – and at less than one-tenth the cost.

Beagle 2 will leapfrog the opposition because it can hitch a lift on board a bigger European mission which is heading off in 2003 to orbit the Red Planet – a mission that owes its existence to the demise of the giant Russian probe Mars 96. As well as its Russian experiments, Mars 96 was carrying a sophisticated suite of European instruments to check out Mars's geology and atmosphere from orbit. Spare copies of these instruments will now survey the planet from Mars Express.

Meanwhile, the tiny Beagle 2 will detach from its host and speed down through Mars's thin atmosphere. Like Pathfinder, Beagle 2 will slow down on parachutes and then bounce on airbags until it reaches its resting place.

The British mission will touch down even before NASA's next landing mission with its pair of rovers

'Pathfinder was a very charismatic mission, a very good PR mission,' says Pillinger, 'because it was the first mission back to Mars for twenty-odd years, and it had the photogenic little rover moving about. But it didn't address the question on everybody's lips: is there life on Mars?'

With Beagle 2, Pillinger combines this ultimate question with a good dose of PR. Early on, he formed a relation with the rock band Blur. Two of the band members, Alex James and Dave Rowntree, are space and astronomy fans, and their 1999 album *13* features a number entitled 'Beagle 2'. This song will be the signal sent home from Mars's surface to indicate that the spacecraft has safely landed.

'This is just one of the situations,' says Pillinger, 'where we can hopefully – by doing something different – attract people to take an interest in science. People who normally wouldn't listen to a science broadcast or watch a science television programme will take an interest because Blur have made science cool.'

Once it's landed, Beagle 2 will extend a robotic arm. At its tip is a platform bearing six scientific instruments splayed out in different directions: the PAW. Pillinger freely admits that 'PAW' doesn't stand for anything: the name derives from its appearance and the canine ancestry of the spacecraft itself!

Two of the 'toes' on the PAW constitute a pair of cameras to provide telescopic views. And – in another move that firmly weds PR and science – the cameras' first images from Mars will include a painting by controversial British artist Damien Hirst.

Pillinger had been wrestling with the problem of how to ensure that ground controllers correctly display the colours sent back by Beagle 2's cameras, when he chanced to watch a TV programme on Hirst. 'This idea was going through our minds that we need an array of targets; and as we watched the programme we equated our array with Damien's spot painting. We contacted Damien, and he thought it was an absolutely tremendous idea that his art could be used for a scientific purpose.'

Now it will be time for Beagle 2 to extend its PAW again, to commence the search for life. One key experiment is to check out the kind of carbon contained in Martian rocks. On Earth, it comes in two stable kinds – carbon-12 and carbon-13 – as well as the radioactive carbon-14 that archaeologists use to tell the date of ancient pieces of wood and bone.

'Biology has a preference for carbon-12,' explains Pillinger. 'If you look through the Earth's sedimentary rocks through the past 4 billion years, you find that the organic matter from living organisms always has 3 per cent or so more of the carbon-12 than the other rocks.'

This carbon-12 test will reveal whether Mars hosted life at any point during its long history – and all without having to winkle out any microscopic fossils. Then Beagle 2 will sniff the air to check if there is still life on Mars today.

'It's the equivalent of looking for the gases that cows produce!' jokes Pillinger. The gas in question is methane. Earth's atmosphere contains a tiny amount of methane, but it disappears quickly as it's attacked by oxygen. 'If it weren't for the biological production of methane on Earth, there wouldn't be any in our atmosphere,' continues Pillinger, 'and a major contributor to this methane is the microflora in the guts of ruminant animals like cows.

Beagle 2 is the leader of a pack of future missions set to challenge Mars's secrets

'Now that isn't predicting cows on Mars, of course,' says Pillinger, 'but it's predicting that somewhere – maybe a thousand kilometres away or a thousand metres down – there could be a hidden population of microflora doing the simplest metabolic process that's known: converting carbon dioxide to methane.'

This conversation brings us to the idea of Gaia, pioneered by British chemist Jim Lovelock, in which the Earth has a single planet-wide living system consisting of all the plants and animals plus the oceans and atmosphere. As part of a living system, the atmosphere won't be an inert mix of gases but will be a mixture of reactive gases, like oxygen and methane.

In 2008, the Mars Smart Lander will use its built-in intelligence to seek out a safe landing spot. Its piggybacked 'roving science laboratory' will check out these possible watery spots, to pave the way for future robotic and manned missions.

Lovelock has analysed the atmosphere of Mars and found no sign of any chemical imbalance that would indicate life on the planet. So why is Pillinger bothering to look?

'There are only six components which are genuinely analysed in the atmosphere,' Pillinger counters. 'All the others might or might not be there. With Beagle 2 we are intending to do a much more detailed analysis of the atmosphere. I know Jim Lovelock rather well, but I don't think he has enough data to be able to make that statement yet.'

Lovelock, unsurprisingly, doesn't agree: 'Well, I would say it's a waste of time. It's just about worth looking for fossil life, but to look for present-day life is flying in the face of the evidence to an extraordinary degree.

'The cost of a Mars expedition is so great,' he continues, 'that it seems a pity to waste energy on a worthless objective, when there are so many worthwhile ones – such as learning about what sort of a planet Mars is, and how it works.'

Pillinger retorts: 'Beagle 2 has an enormous amount of science. We're not just sticking with life. Our experiments that study the atmosphere will tell us about its composition, and hence the history of Mars and about climate changes over the seasons on Mars. And our experiments to analyse the rocks will tell us things like their age and how long they've been in the places where we find them.'

Beagle 2 is the leader of a pack of future missions set to challenge Mars's secrets. Appropriately, they cut across the Earth's national boundaries. As the European Mars Express orbits the Red Planet, it will be joined in 2004 by the Japanese spaceprobe Nozomi, which will make a pioneering study of Mars's mysterious upper atmosphere. While the Earth's blanket of air is largely protected by our planet's magnetic field, Mars's atmosphere is exposed to the full force of electric storms sweeping out from the Sun.

The Russians hope to return later, in their long-standing quest to understand Mars's two small moons, Phobos and Deimos. French researchers have experimented with huge balloons that could loft heavy payloads over the red deserts. An international network of dedicated satellites eventually will enable robots to communicate more effectively with each other and with the Earth – providing a Mars-Internet.

Mars rover pioneer Rob Manning sees the beginning of this development already. 'The Mars Express UHF radio that can communicate with Beagle 2 on the surface could be useful for future US missions, and vice versa. US missions will be able to support Beagle, potentially, or other surface missions – there's a lot of international cooperation at that level.'

'There is no doubt that the Mars programme is an international programme,' says NASA's 'Mars Tsar', Scott Hubbard. 'It helps to bring in other elements where the whole is hopefully greater than the sum of the parts.'

And, as Manning reminds us, an international approach is – in the long term – the only appropriate way forward: Earth to Mars, planet to planet. Standing in front of a poster of Pathfinder and the Sojourner rover – signed by all the team members – Manning takes a sweeping view of the past and the future. 'You know, these are sitting among rocks that haven't changed since Mars was young. Apart from the spacecraft, this scene has remained unchanged for billions of years.

'Now suppose human civilisation were to die out in 20 to 30 million years. Then in, say, 200 million years all traces of human civilisation will be wiped off the Earth, by erosion and plate tectonics. So long after there's no trace that humans ever existed on Earth, these guys on Mars will still be there, looking almost identical to how they do now.'

MARS ON EARTH

CHAPTER 9

> *The most Mars-like place I've ever been is a region in Antarctica called the Dry Valleys. It's got a mean annual temperature that's about twenty-five below zero. There's no vegetation whatsoever – it's very cold, very dry, very windy. When you walk around, it looks superficially very much like Mars.*

Planetary geologist Nathalie Cabrol will never forget the day when man landed on the Moon. It completely changed her life. 'I was five years old when Armstrong stepped on the Moon, and that really did something about my way of thinking. I said, "OK, I want to be an astronomer, but in the meantime I want to explore the planets". When I was sixteen I applied to be an astronaut in France, but I was too young at the time.'

Cabrol, now based at the Ames Research Center in Silicon Valley, is a vivacious and passionate Frenchwoman who lives, eats and breathes planetary exploration. Even the photograph on her website sees her kitted out in an orange spacesuit. She is driven to explore. 'It's as if I was a little girl sitting in a field in summer, and watching the stars,' she recalls.'It was beautiful. But there was something else. I wanted to understand where we are coming from, but I also wanted to touch – but to touch a star is impossible, you know.

'And so this was my way to try to put together my passion for astronomy, and my passion for the beauty of landscapes. I guess planetary geology is a mix of all that. And then Mars came along, because Mars is a very terrestrial-like planet. And a sunset on Mars is something.'

Despite its temperature of –50°C, Don Juan Lake in Antarctica is liquid, owing to the high concentration of salts dissolved in the water. It is also teeming with microbial life.

But how can Cabrol touch Mars? 'You have lots of landscapes on Earth that remind you of Mars. Since we cannot go to Mars right now, you have to find what we call analogues here on Earth. Of course, the Earth will never be Mars, so a perfect analogue does not exist. So you have to find bits and pieces all over the place.

'Sometimes you want to find an analogue for what conditions on Mars were like in the past, so you will go to ancient dry lake beds in the desert. Or you might need an analogue which is good for the climate – then you will go to the polar regions to find low temperatures and the work of ice melting.'

'The most Mars-like place I've ever been is a region in Antarctica called the Dry Valleys,' enthuses fellow planetary geologist Steve Squyres. 'It's got a mean annual temperature that's about twenty-five below zero. There's no vegetation whatsoever – it's very cold, very dry, very windy. When you walk around, it looks superficially very much like Mars.'

Nathalie Cabrol's Ames colleague Chris McKay – affectionately nicknamed 'Mr Mars' because he is so preoccupied with the Red Planet – cut his research teeth in the Antarctic Dry Valleys. 'I ended up getting involved with a team that was going down to Antarctica to study life, and the Dry Valleys that were the Martian analogue. It was an important step for me, because it got me directly involved in studies of life and extreme environments as related to Mars – and essentially I've been doing that ever since.'

McKay sits in a cluttered office at NASA's Ames surrounded by finds from previous field trips. From his desk, we pick up a rock that once lay on the Antarctic ice sheet. 'There are microbes inside that rock,' he points out. 'It's surprising the places we find life. The most surprising thing is how clever life is in getting into the essential environments

'Since we cannot go to Mars right now, you have to find analogues here on Earth'

A photomicrograph of a colony of cyanobacteria, which thrive at the frozen South Pole. This particular variety, *Gloeotrichia echinulata*, has numerous filaments composed of many individual cells.

that it needs. And the thing that characterises the ecosystems we're studying in all the different environments is the presence of liquid water. The organisms are just more clever at getting the liquid water than we could ever have imagined.'

Although Earth's hot deserts may bear a passing resemblance to the surface of Mars, and their dry lake beds are excellent analogues for the Red Planet's ancient watercourses, it is to polar climes that researchers trek today for a taste of Mars on Earth.

David Wynn-Williams is a microbiologist for the British Antarctic Survey in Cambridge, and he takes every opportunity to get down to the South Pole.

'The Antarctic deserts are ideal analogues of Mars – I rest my case,' he maintains. 'Not only do you have the dry soils, but the ice-covered lakes. So if you're familiar with Chris McKay's model for the evolution of water on Mars –

'The most surprising thing is how clever life is in getting into the essential environments that it needs' which starts off with water everywhere – then obviously you could theoretically get life. The water then recedes into ice-covered lakes, which are probably hypersaline, which we also have in the Antarctic. Then it would recede into rocks which contain moisture, and eventually you'd end up with a desert. This is what we have on Mars at the moment, and it's also what we find in the Antarctic.'

> **Antarctica teems with microbial life, despite the freezing temperatures and lack of sunlight**

Wynn-Williams and his colleagues never feel far from Mars when they go on their polar expeditions. 'By an amazing coincidence, the area of Alexander Island where we work – which is about 72° south – is full of astronomical names. When we set up this new cold desert research site, it was alongside the Mars Glacier, where we camped. We named the little valley we were working in Viking Valley. Now we have another site just round the corner called Mars Oasis, and we camp on the Utopia Glacier, which I named after the site where the Viking 2 lander came down. I'm waiting for the opportunity to write the ultimate article, when we have some of our equipment on Mars. It'll be called "From Mars Oasis to Mars".'

Antarctica teems with microbial life, despite the freezing temperatures and lack of sunlight. David Wynn-Williams is particularly interested in cyanobacteria – organisms like blue-green algae which use sunlight to produce energy. They evolved very early in the history of the Earth, some 3.5 billion years ago, and Wynn-Williams believes that they might have got started on Mars. 'If there was water on Mars and there was light – and a certain amount of organic material – then you've got the melting pot for cyanobacteria to develop.'

Wynn-Williams finds cyanobacteria in two main locations at the southern icecap. One is at the bottom of deep, ice-covered lakes. Even though the ice cover is many yards thick, the water is very clear and sunlight can penetrate all the way down to the bottom. Layers of cyanobacteria alternate with sediments washed down from the surface, forming peculiar animal–mineral structures called stromatolites.

Rocks – like the sample in Chris McKay's office – are the other favoured location. 'In the Transantarctic Mountains, you get a lot of sandstone,' explains Wynn-Williams. 'Light can penetrate into this as much as eight millimetres. So as you crack the rock open, you can see these cyanobacteria in layers, a bit like a liquorice allsort.

'The innermost layer is receiving just one-thousandth the light it would receive on the surface, and they can live quite happily in that environment. So bacteria like these could well have been the last microbes to have survived on the surface of Mars.'

The South Pole is also Mars-like in another respect. 'It has the ozone hole,' continues Wynn-Williams. 'We get massive amounts of extra UVB because there is very little protection during the winter. So we're also studying the response of cyanobacteria to extreme UV.'

Antarctica is not the only polar analogue for Mars. In the Arctic Circle north of Canada lies Devon Island – at 300 miles in length, the world's largest uninhabited island. It is the latest mecca for those seeking Mars on Earth, and the scene of a daring and ambitious experiment that will pave the way for human exploration of the Red Planet. We mention Devon Island to David Wynn-Williams. How does it compare with Antarctica? 'It's very interesting you should mention that, because we have a guy who goes to both. That's Charlie Cockell.'

> 'By the end of the evening, we essentially had a plan to set up the Forward to Mars Party and run in the election'

The Harrow-educated scientist follows in the tradition of those nineteenth-century British explorers who reached all four corners of the globe. In 1990 he organised England's first expedition to Mongolia since 1921, and followed this in 1993 with a moth-collecting trip to the Indonesian rainforest using a revolutionary microlight aircraft. And all this before turning thirty.

Charlie Cockell has already made it into the history books. In 1990 – aged twenty-three – he stood against the then British Prime Minister John Major in the General Election. He explains what led to this extraordinary turn of events. 'I was doing my doctorate in molecular biology at Oxford at the time. One night, I was sitting in Corpus Christi bar, and I was talking to some people about the exploration of Mars. And I was saying how the government should take a greater interest in the European Space Agency and in Mars missions.

'And my friend turned round and said, "If you feel so strongly about this, why don't you stand in the General Election?" – it was about two months off at that point. I sort of laughed, but as we got talking it got out of hand, and the people with me volunteered to be Shadow Environment Secretaries and Shadow Education Secretaries. By the end of the evening, we essentially had a plan to set up the Forward to Mars Party and run in the election.

'Before we knew what had happened, it had got a life of its own. We went to the Prime Minister's constituency – Huntingdon – every weekend, and canvassed. I stood on the litter bin in the High Street and gave lectures on the exploration of Mars while people were walking

Watercolour of Cape Wadsworth on Coulman Island, Antarctica, by Edwin Adrian Wilson (1872–1912). As well as its cold climate, the island consists of overlapping shield volcanoes – another similarity with Mars.

'All these craters have unique microbiology – from cyanobacteria eking out a living in rocks up in the polar regions, to prolific microbial mats you find in the Pretoria Salt Pan'

out of Tesco with their shopping. We got a really good response.

'We had one problem. We were up against Lord Buckethead, who, as you may or may not know, claims to come from the planet Gremloid – he's a complete nutter. On their planet, they speak with a ball of silly putty, which they pull into funny shapes. So people would ask him things like, "What are you going to do about milk deliveries in Huntingdon?" And he'd sit there for ten minutes pulling his ball of silly putty in complete silence. And then his translator would explain what it all meant in English.

North Polar Base by Michael Carroll (1984) was the first depiction of a polar base on Mars. The explorers are taking samples for research with a dirigible (right), while those on the steep hillside (left) check communi-cations links.

'So it was quite an interesting election campaign. We got ninety-one votes in the end, beating the Natural Law Party with about fifty. But Lord Buckethead got 110.'

Did the Prime Minister venture any opinion? 'I had quite a good talk with him about space exploration on election night, before they read out the votes. He was quite enthusiastic about it. But he went along with what was then his Party's line, which was that it's OK for Britain to put money into Earth observation but didn't see any reason to be involved in Mars exploration.'

Charlie Cockell has been fascinated by Mars since the age of seven. 'There was a book at school about the human exploration of space, and the last chapter was all about Mars bases. It described how we would be on Mars in the 1980s, with permanently occupied bases. And I worked out that by the time I was in my twenties there might be a possibility of going to Mars.'

But by 1991, when Cockell attended the International Space University in Toulouse, humankind was no closer to landing on Mars than it had been in the 1970s. Cockell adjusted his sights and decided to aim for NASA instead. 'The design project that summer was the International Mars Mission. And I thought, This is the ideal opportunity to try and make this dream of working for NASA come true.

'I went there and had a really good summer, and met Chris McKay. I said, "Look, I've always wanted to work at NASA". They said, "Why don't you come over and do a post-doc?" which I did the next term. And that was how I got set up with a National Research Council Associateship and ended up at NASA.'

Cockell's work at NASA focuses on the biology of impact craters. Although only about 150 have been identified on Earth, they contain a very wide range of ecosystems. They range from craters in rainforests, one in South Africa with bushveldt vegetation, to the Pretoria Salt Pan, which is filled with a hypersaline lake. 'There's one in India that's inhabited by crocodiles,' Cockell enthuses, 'and all these craters have unique microbiology – from cyanobacteria eking out a living in rocks up in the polar regions, to prolific microbial mats you find in the Pretoria Salt Pan.'

But the crater that preoccupies Cockell at the moment is Haughton Crater, in the Arctic wastes of Canada's Devon Island. He's grateful to his NASA-Ames colleague Pascal Lee for introducing him to both the crater and to the island. 'Pascal did a post-doc to study the Haughton impact crater up in the Arctic. And – in a sense – I owe him for my interest in impact craters. He set up the NASA project there, and was looking for a biologist to lead the exobiology effort, which I ended up doing. And the project has grown from just a few people in the first year to being a really major project. It's got all these angles of biology, but also communications and logistics – the background logistic needs for planning a human mission to Mars.'

Pascal Lee – planetary geologist and Federal Aviation Authority-certified helicopter flight instructor – heads up the Haughton-Mars Project (HMP) on Devon Island, which he initiated in 1997.

'In my view,' he reflects, 'there is really no place on Earth that is like Mars. Mars has all these alien conditions – a 0.38 gravity, an atmosphere as thin as the Earth's at about 100,000 feet, the high UV radiation environment at the surface, and very low temperatures. It's a non-Earth place.

'But there are places on our own planet which approximate in one way or another to some aspects of Mars, and that's certainly the case on Devon Island.'

Spacesuited Pascal Lee – rock hammer and map in hand – explores the 'Martian' landscape in Haughton Crater, Devon Island.

To Pascal Lee, Devon Island is far more than just a Mars analogue. 'The place is absolutely incredible. It's sort of interesting that nothingness can stir you so much – it's a bit like deep space, I guess. Here you have a landscape that's almost devoid of any vegetation; it's bleak, it's barren, it's a relatively uniform colour. It really has nothing – no power lines, no signs of civilisation – and yet it has this mesmerising appeal. I would say it has a primordial beauty to it.

'Here you have a landscape that's almost devoid of any vegetation; it's bleak, it's barren, it's a relatively uniform colour...and yet it has this mesmerising appeal'

'In a very profound sense, when you're there, you have this feeling of being on another world. The people who have been there are so few that, in all likelihood, every step you make is pretty much the first one that's ever been made there. You're faced with this powerful sense of actually exploring the place and discovering it for the first time. It's something we don't really have a chance to experience any more on Earth very much.'

Charlie Cockell has spent four summers on Devon Island. 'It can be really pleasant.

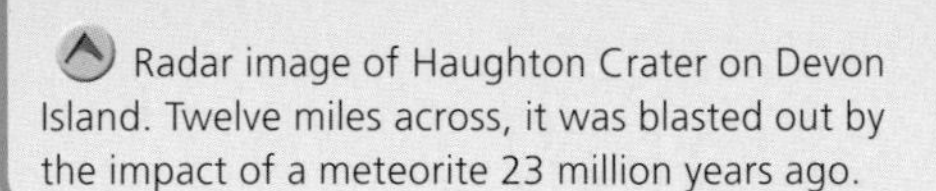

Radar image of Haughton Crater on Devon Island. Twelve miles across, it was blasted out by the impact of a meteorite 23 million years ago.

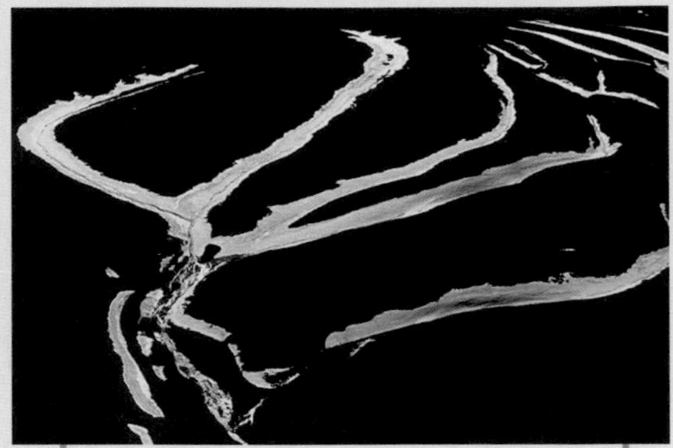

Aerial view of long, finger-shaped valleys near Haughton Crater, which resemble channels on Mars.

Then you get these cold fronts that come in, and it's freezing. You get howling winds up to about fifty knots that rip through the campsite, so it changes quite rapidly. It's very, very dusty – this is an area where it's much like Mars, really. There's a thin dust layer that gets carried up into the atmosphere during the winds, and it gets absolutely everywhere – which, of course, would be one of the major problems on Mars.'

Geologically, Devon Island bears a distinct resemblance to Mars. It's dissected by networks of finger-like valleys and canyons which are very strange by terrestrial standards. Pascal Lee believes they are related to their counterparts on the Red Planet. 'We feel there's a common history. It all ties into a very cold past on Mars – not an early Mars that would have been warm and wet, as it's classically pictured, but a planet that had a cold climate all along. We need to work out the details as to how valleys and canyons form in the cold – it's associated with ice covers, ice sheets and glaciers.'

The twelve-mile-wide Haughton Crater, which drew both Lee and Cockell to Devon Island, is named after the Reverend Samuel Haughton, a Victorian naturalist who wrote the first account of the geology of the Arctic. It was blasted out 23 million years ago by the impact of a large meteorite. The only impact crater on Earth in a polar location, it is an irresistible analogue for the frozen craters on Mars. Charlie Cockell – passionate about Mars and impact craters alike – waxes lyrical about it.

'Impact craters are sort of like nature's drill,' he enthuses. 'You can use impact craters to drill into the subsurface of Mars in the way people have suggested deep-drilling to uncover a buried biosphere. I think it's the best way for looking for life on Mars, because you get ponding of water in these ancient craters. They may have been a transient habitat for life – particularly if you have impact heating as well, which might provide a thermal path for microbes.'

'Haughton Crater itself hosted a lake,' adds Pascal Lee. 'The waters are long gone, but the lake sediments are still there. And you find in them a beautiful environmental record of the site – the climate, and the flora and fauna that used to live here. They're preserved in an amazing state: rhinoceros and rabbit bones that aren't even petrified. You can pick up wood from 23 million years ago that's still burnable – it's all been preserved in the permafrost in the lake bed. And that's giving us a lot of hope that we'll find the same sort of things on Mars.'

The NASA scientists do not have the magnificent desolation of Devon Island all to themselves, however. They share their summers on Devon with people who – on the face of it – might appear to be strange bedfellows. These are members of the Mars Society – a pressure group which advocates the human exploration of Mars as the next goal in space.

'NASA is suspicious of the Mars Society, so it's politically difficult,' admits Charlie Cockell. 'The Mars Society believes very much in free access to Mars for all people, rather than have it be a government-controlled activity. So there are two very different views on how you explore Mars here, and I think it's probably healthy that they're working together.'

Pascal Lee was responsible for the unlikely marriage. 'There are two projects on Devon Island that are taking place. One is motivated by the scientific appeal of the site, and we've been going there for the past four years to do, above all, geology and biology. But since year two we've been developing an exploration programme – learning how to explore Mars with humans and robots.

'I was one of the founding members of the Mars Society, so I invited them to come and test out their hardware, to learn how to explore with us, so we can sort of do some of the homework that's going to be needed to send humans to Mars.'

Chris 'Mr Mars' McKay initially sowed the seed which grew into the Mars Society. 'We were a bunch of graduate students who got together and saw that Mars was potentially a very interesting place, where humans could go and explore it for evidence of past life. There were no official memberships or party line or anything – it was very open and fluid, and people from many different perspectives could come and share their views, and that was part of the strength of it. Because we were just graduate students, a reporter from Washington described us as "the Mars Underground" – and it stuck.'

'It became clear that we're better prepared to send people to Mars than to send people to the Moon'

The Mars Underground movement quickly surfaced. 'It became the opposite of an underground,' recalls McKay. 'People like me went on to be researchers in NASA centres and universities. Mars became less problematical a topic to discuss. And the Mars Society has taken over the public outreach and advocacy that was part of the Mars Underground.'

The man behind the Mars Society is Bob Zubrin. 'As an engineer in the early 1990s, I was responsible for developing plans for Mars missions. It became clear that we're better prepared to send people to Mars than to send people to the Moon. For example, in 1961 – if we'd been serious and had taken an intelligent approach – we could have had people on Mars within ten years.

'I wrote a book outlining that plan, and I got over 4,000 letters from people all over the world, saying, "How do we make this happen?" And I discussed this with friends like Chris McKay, and we decided that the time was right to pull together a Mars Society – an international society

Teams from NASA and the Mars Society begin to erect the Habitat on Devon Island. Even visiting journalists were roped in to provide muscle-power.

committed to making a reality of getting humans to Mars. We had our first convention in Boulder in August 1998, and 700 people showed up from 40 countries.'

Bob Zubrin isn't in residence when we turn up at his aerospace company in Denver in July. He's away on Devon Island – leaving the running of the Mars Society to his wife, Maggie. 'I'm missing Bob, and my sleep patterns are going crazy,' she explains. 'Bob's helping erect our Habitat, and as soon as that's up, he will turn to research.'

Pascal Lee was the prime mover here. 'I proposed to the Mars Society that its first project ought to be the establishment of a Habitat on Devon Island – one that would be shaped like a landed spacecraft on Mars, which would give an additional level of realism to the simulation we'd like to do. NASA doesn't have a presidential mandate or congressional approval to do a mission to Mars, but if we could find private funds to put together a Habitat, it could serve as a simulation test-bed for us.'

But things didn't exactly go according to plan. 'I don't know if you're completely aware,' explains Maggie Zubrin, 'but we did lose our crane, our floor, and our trailer. So the first thing we had to do was line up an alternative construction system, which is essentially manpower. And in fact, if you had joined him on Devon, you would probably be lifting panels yourselves.'

'It was quite an adventure,' recalls Bob Zubrin. 'There are no big airfields on Devon Island, other than bits of flat land that little twin Otter aircraft can land on. So to deliver the Habitat, we had to do it in the form of para drops. The first four para drops made it to the ground safely, but the fifth and last para drop was a complete disaster. The payload separated from the parachute at an altitude of a thousand feet, and then hit the ground at some 400 mph. The payload was completely destroyed – and in it were the advanced fibreglass floors for the Habitat.

'Then the crane that was needed to erect the wall panels was destroyed as well. And the construction crew that had flown up to build the Habitat then fled the island. At this point, most people thought we were out for the count – in fact, I was interviewed by a reporter who said to me: "Dr Zubrin, how would you compare the failure of your mission to that of the Mars Polar Lander?" And my answer to him was that there's a parallel – we've both hit a rock – but the difference is that we have a team on the scene here, and we're going to find a way out of this.

'And that's exactly what we did. We found wood in Resolute Bay to replace the floors. We also built the new trailer out of wood and out of parts of a wrecked baggage cart from Resolute Bay Airport. It was a very strange contraption, and it constantly fell apart. But we kept nailing it back together, and with it we were able to move the heavy wall panels from where they had fallen up to the ridge.

'We pulled together a new construction team out of a mixed group of scientists and journalists, who I pressed into service using what might be called ancient Roman construction techniques. We had large labour gangs working with block and tackle and scaffolding and so on, and we got the whole thing up.

'There's a lesson here, which is – while the particular experience on Devon Island would not be the same as that on Mars – the fundamental principle is the same. When machines break, it's humans who find a way round. On a human Mars mission, the crew is going to be the strongest link in the chain.'

Charlie Cockell is well aware of this, and soon he's going to be able to put it to the test. When we spoke to him, he could hardly contain his excitement. He's off to Houston in a few days to discuss linking up Mission Control with Devon Island for a dress rehearsal of a real human Mars mission. 'This summer we're going to wear simulated spacesuits to go out and do science exercises like collecting rocks and looking for microbes. We want to find out how the logistics of being on Mars actually constrains the research that you do. How does a spacesuit limit your ability to work in the field? How do the communications delays affect the way you do science?

'Mission Control in Houston will set themselves up as if it really were a real Mars mission and will introduce a twenty-minute time delay into our communications and see how that affects the pattern of research. On the Moon, you can communicate in real time, and every second of the time the astronauts spent on the Moon was absolutely planned – where they were going to go, what rocks they were going to pick up – nothing changed without recourse to Mission Control.

'When machines break, it's humans who find a way round. On a human Mars mission, the crew is going to be the strongest link in the chain'

'Now on Mars, you've got a stay of nine months, you've got a twenty-minute time delay, and there's no way you can have constant control from Houston telling you what to do. So what we want to know is: should Mission Control become Mission Advice?'

If science, rather than logistics, sets the agenda, then NASA will have to face up to a new dynamic when it comes to the first real human Mars mission. 'It's very controversial,' admits Cockell. 'It affects people's jobs in Houston. There are really two camps on this, and they vehemently fight it out. There are those who say, "Absolutely not – everything has to be planned." Then there are other people who say that there has to be a complete sea-change.'

No doubt Charlie Cockell – explorer, scientist, leader and politician – will be lobbying fiercely on the side of science when that time comes.

DESTINATION MARS

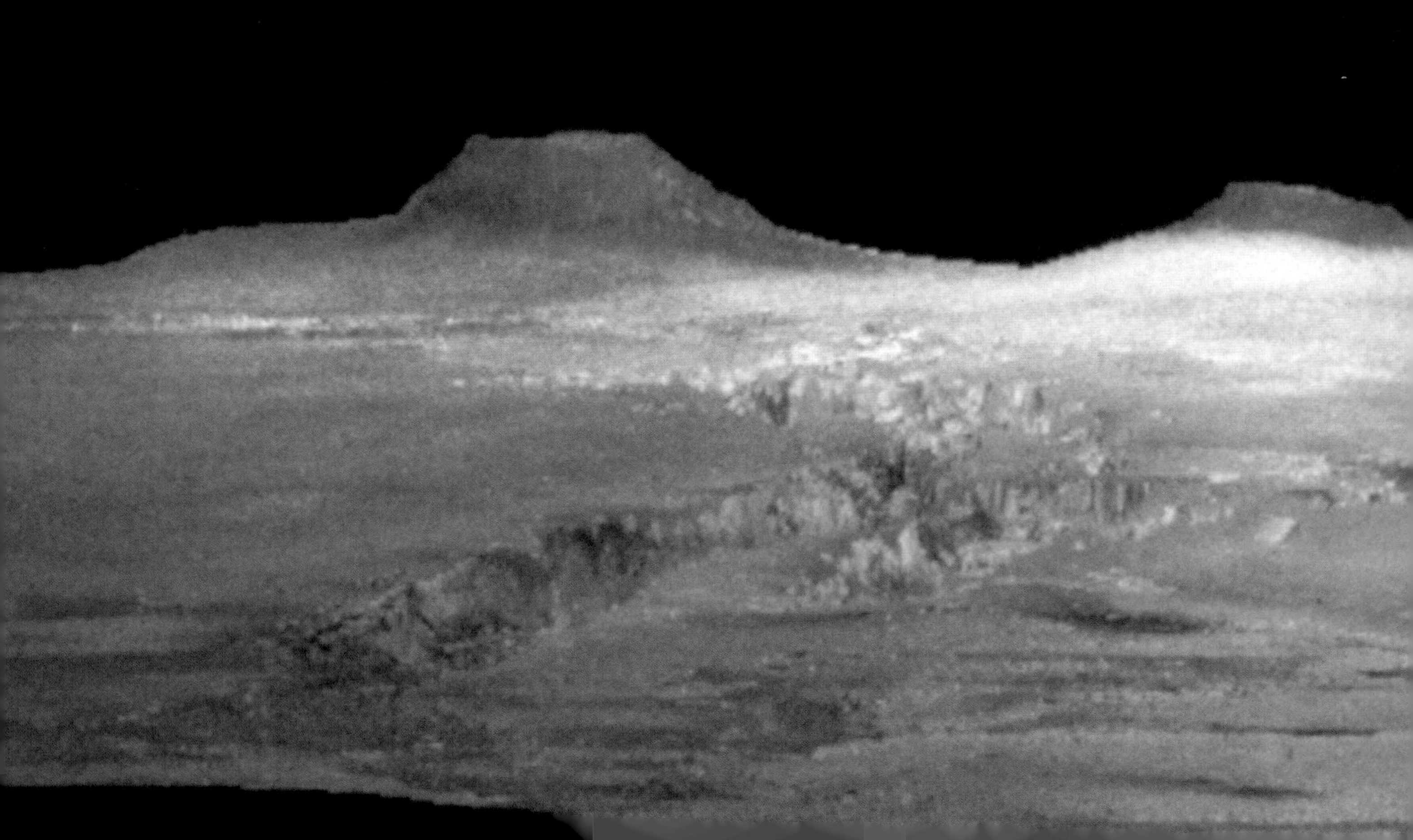

CHAPTER 10

'Within our lifetimes and within our careers,' asserts Hum Mandell. 'You know, this is the frustrating part about working on this, because I think all of us in this room believe that we could begin sensibly in the next few years. Within a decade, we could do the job.'

The folk at NASA's Exploration Office in Houston are in the mood for telling it like it is. 'I'm Hum Mandell – Hum is short for Humboldt. We're convinced that the reason humans aren't going to Mars doesn't have much to do with engineering. It has more to do with politics.'

So when will the first humans get to Mars? 'Within our lifetimes and within our careers,' asserts Mandell. 'You know, this is the frustrating part about working on this, because I think all of us in this room believe that we could begin sensibly in the next few years. Within a decade, we could do the job.'

The International Space Station: a training-ground for long-duration spaceflight, a stage on which to rehearse a human Mars mission – and the ultimate tourist destination.

Until this moment, the thought that had been uppermost in our thoughts was when *we* might get to the Exploration Office – let alone Mars. A call to an astronaut friend fixes things for us. One of his most respected colleagues is Brenda Ward, manager of the Exploration Office. She is not only willing to talk, but rallies around nearly a dozen of her colleagues too. In a hectic conference call, we pick up some of the excitement and vision that are coming out of one of the most creative teams in NASA. Its focus is on future human space exploration in all its shapes and forms – and Mars is part of the agenda.

'I'm Joyce Carpenter. My main focus is on our Advanced Design Team, who are actually coming up with concepts for the vehicles and habitats that we will some day send to Mars.'

'John Connolly. I'm one of the engineers here, and my main concern is the transition between Mars robotic missions into the human mission.'

'My name is Jeff George. I'm one of the handful of people here trying to lay out these missions. I guess my speciality is advanced power and propulsion systems: nuclear power, nuclear propulsion, electric propulsion – the different ways we might generate power and actually propel ourselves to Mars and back.'

Not surprisingly, the team has close connections with the astronaut corps, who are also based at Houston. 'We enjoy talking about these missions,' smiles Doug Cooke, who is the overall head of the Exploration Office. 'John Young is one in particular – he's a real advocate. We include them in meetings or in our discussions so they can make inputs.'

Several of the team made applications to become astronauts, but discovered how fierce the competition was. 'We all realise that astronauts are sort of top of the pyramid,' observes John Connolly. 'But really, the reason why most of us work here is because we want to be part of the adventure. The closer you get to the top of the pyramid – that's great – but just being part of the pyramid is worth it.'

But there is a fly in the ointment. How definite are NASA's plans to send a human mission to Mars? The Exploration Office team suddenly becomes rather more subdued. 'Not as definite as we'd like,' says Hum Mandell ruefully. 'Officially, NASA has no plans to send people to Mars right now,' comments John Connolly.

'The kind of work we do is more in the arena of strategic planning,' explains Doug Cooke. 'We are also working with the robotic community. In fact, I represent the human exploration

So when will the first humans get to Mars?

and development of space enterprise at NASA Headquarters to Scott Hubbard.'

'We don't have a timetable for the human exploration of Mars,' admits 'Mars Tsar' Scott Hubbard. But he makes it clear that – even as a very senior NASA official – his hands were ultimately tied by Congress. And the mandate of Congress is to finish the International Space Station and demonstrate that it works before taking on human exploration projects beyond low Earth-orbit.

A small unmanned Scout plane that NASA is planning to send to Mars, possibly as early as 2007. In order to fly in the thin Martian atmosphere, the plane needs enormously wide wings.

Hubbard is aware that not everyone shares the view of Congress. 'There are many people in NASA and throughout the world who are interested in the human exploration of Mars. I get e-mail all the time from people who think this should be a priority.

'But I think there are four bits of homework that are required before we do it. One is the big why – what is the driving reason to explore Mars? My answer, as Scott Hubbard, would be that – of all the planets we could get to easily – it's the one most likely to have harboured life, or maybe have life today under the surface. Second is that we need to have the technology to do this. Some pieces are probably ready, while others need investment – whether it's advanced life support or propulsion, whatever.

'Third is that all this needs to be affordable. And then the fourth thing is that we need to understand – and where necessary address – the biological challenges of sending humans out on a mission that could be as long as three years out and back, in a radiation and microgravity environment that we don't understand very well today. So that's our homework assignment for sending humans to Mars.'

Part of the assignment will be seeing how things pan out on the International Space Station, which is currently under construction. 'It'll give us a tremendous amount of information related to the biomedical and biological aspects of long-term space travel – that was one of its reasons for its existence,' explains Hubbard.

'There are many people in NASA and throughout the world who are interested in the human exploration of Mars'

The other way ahead is through robotic exploration. 'I interact directly with Scott and the Mars Robotic Programme,' says Doug Cooke. 'That's because the robotic programme holds a lot of potential for us to learn about the environment that we send people into when they go to Mars.

A few scientists even believe robots will become so clever that they will remove the imperative to send humans to Mars altogether. Mike Carr, the geologist who so enthusiastically scanned the surface of Mars daily for a safe landing site for Viking 1, is perplexed by the notion of sending people there. 'I'm not a real strong proponent for human exploration,' he admits. 'I have a little trouble understanding the rationale.'

But surely – as a geologist – wouldn't he love to pick up Martian rocks, explore the planet's vast canyons and climb its enormous volcanoes? 'There are lava flows going over the huge cliffs around Olympus Mons, and so – yeah – it would be exciting to see Mars. But I bet, in the next ten, fifteen years we're going to have such high-resolution images from rovers and so forth that we're going to be almost there.'

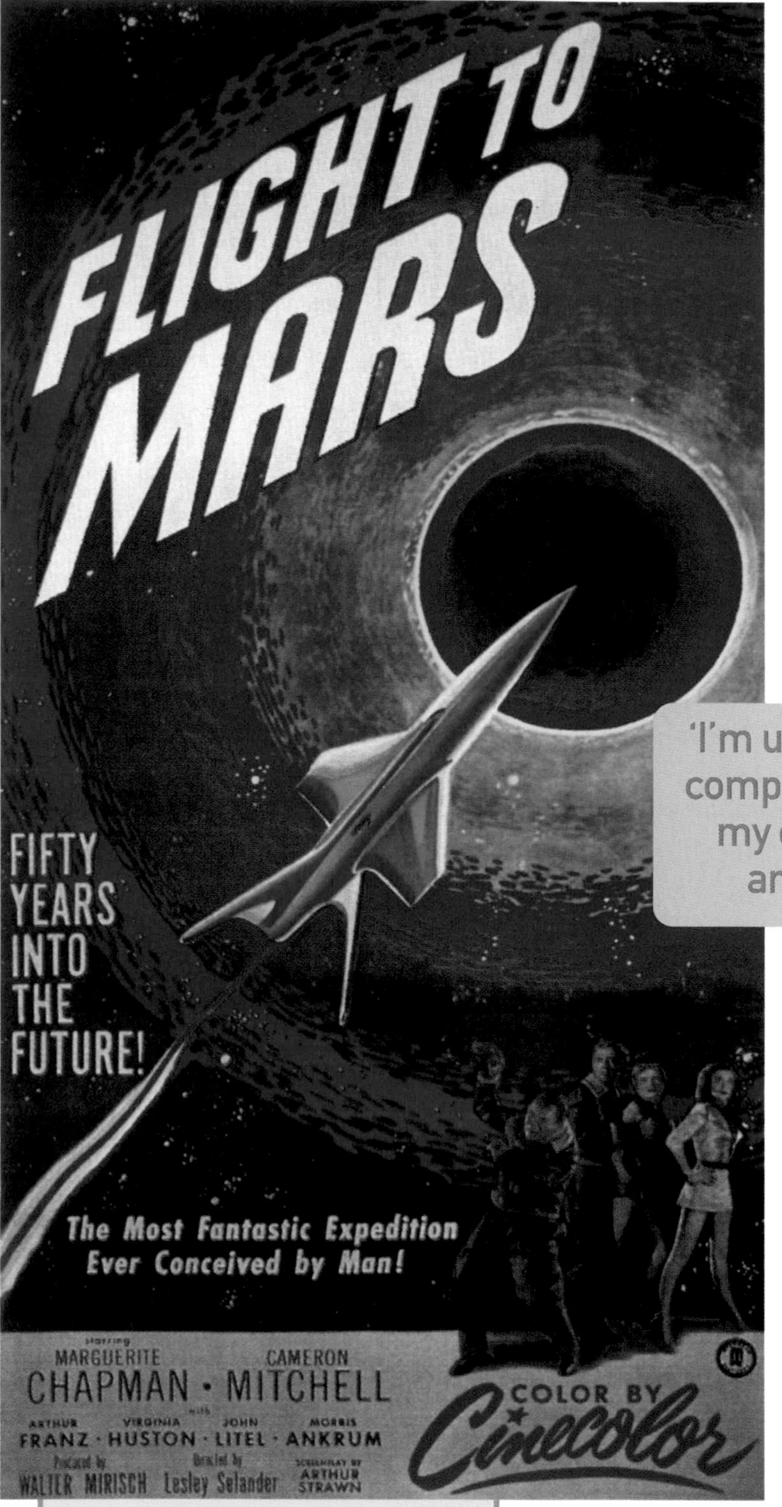

The 1951 film *Flight to Mars* (directed by Lesley Selander) featured a human crew experiencing a meteor storm en route to the Red Planet. The rocket was re-used in a later movie.

'I'm using the most sophisticated computer I have, and it's between my ears. It's much better than any robot we could create'

Fellow geologist David Wynn-Williams disagrees. 'One of the reasons for someone like myself going to Mars would be that I can walk up to a cliff in Antarctica, take one look at it, and say, "Yes, those are the right kinds of rocks to contain microbes." I'm using this remarkable power of vision I've been gifted with, I'm using the most sophisticated computer I have, and it's between my ears. It's much better than any robot we could create. Even the most sophisticated robot is stupid compared with the most dense human being.'

Many people, like sci-fi author Stephen Baxter – whose novel *Voyage* chronicles a human flight to Mars only a few years after the Apollo landings – are disappointed that it didn't all happen earlier. Historian Steve Dick has a lot of sympathy. 'When I was a youngster, and people were heading for the Moon on the Apollo missions, they were projecting that by 1984 we would have a manned landing on Mars. I hope that – in the background – NASA does have an agenda to send human beings to Mars.'

Don't tell Congress, but we found it everywhere we went on NASA territory. John Connolly, from the Exploration Office, revealed that the most recent probe to head off for Mars – the 2001 Mars Odyssey – contains an experiment that is truly looking to the future. 'It has an instrument on board that basically came from this office. It's a radiation monitor – the first instrument flown to Mars for no reason other than to gather some data that's pertinent to future human missions.'

But before humans go to Mars, we may first have to return to the Moon. The Exploration Office is keenly aware that operating in low Earth-orbit on the International Space Station raises few of the demands of a Martian expedition. The team even has plans to fly to points further afield – like the nearby asteroids – to test out the technologies and systems that will be required for exploration far from home. 'When you're up in the Space Station you can get resupply modules, and – in an absolute emergency – you can just get in your escape capsule and come home,' points out astronaut Jeff Hoffman. 'Once you've fired your engines and left Earth on your way to Mars, there's no turning back. You're on your way, and you're on your own.'

People are also under no illusions that the first human flight to Mars will be routine. Veteran space commentator and former Mission Controller Jim Oberg confesses: 'I've come grudgingly to the conclusion that it's not nearly as easy as I once hoped it was. The kind of equipment, the lifetimes required, and the operating environments are so alien.' His concerns are echoed by astrobiologist Chris Chyba: 'My personal prejudice here is that it's going to turn out a lot harder to have human spaceflight of years-long duration between the planets than we appreciate right now.'

'Once you've fired your engines and left Earth on your way to Mars, there's no turning back. You're on your way, and you're on your own'

The team at the Exploration Office has given a lot of thought to the shape of the first human mission. One of the major concerns is to cut down on weight. Having a crew with all their provisions on board – plus a landing craft and a return vehicle – would require an enormous launcher. So the name of the game is 'pre-deployment' – sending vehicles and supplies ahead of the actual voyage, so that everything is in place when the crew eventually arrives.

'Actually, our mission mode right now is in a lot of respects similar to Apollo,' explains Bret Drake. 'We pre-deploy the lander that the crew would use for landing and returning to orbit. Then the crew travels to Mars in a separate vehicle; they rendezvous with the lander in Mars-orbit, then leave their transfer vehicle orbiting Mars while they descend to the surface. Some of our missions – like the long-stay mission, which is a year-and-a-half – are too long to leave a person up in Mars-orbit, so the only difference is the whole crew goes down to the surface of Mars.'

The plan bears a remarkable resemblance to a scheme suggested by Bob Zubrin, now president of the Mars Society. When he was an engineer at Martin Marietta Astronautics (now Lockheed Martin), he and his colleagues were flabbergasted by NASA's estimated price-tag for

Artist's impression of a human outpost on Mars. The crew's ascent vehicle and propellant production plant lie half a mile away. (Although this image was commissioned by NASA, the Agency says in its caption material: 'NASA currently has no formal plans for a human expedition to Mars or the Moon.')

a Mars mission. Zubrin recalls the trigger for his plan. 'Back in 1989, on the twentieth anniversary of the Moon landing, President Bush announced his space exploration initiative – humans to the Moon, on to Mars, this time to stay. And then NASA went on to conduct a study on how this might be accomplished. They came up with an incredibly elaborate thirty-year plan for building giant orbiting space ports that could be used to construct giant interplanetary spaceships.

'Mars has the resources needed to support life. The same resources that make it interesting should make it attainable'

'These spaceships could then be sailed off to Mars fully loaded with all the fuel and oxygen needed for a mission. But in fact seventy-five per cent of what they were shipping to Mars were the fuel and oxygen required to come home. The total cost was in the order of $400 billion, and that immediately killed the programme – there was no way it was going to happen as far as the US Congress was concerned.

'The plan made no sense to me, because it's not how we've explored the Earth. When we've done it intelligently, it's been by living off the land. Mars has the resources needed to support life. The same resources that make it interesting should make it attainable.'

Instead, Zubrin came up with a plan of his own. It would involve two comparatively small spacecraft being launched direct to Mars. The first one flies to Mars unfuelled. After landing, it sucks in Martian air, which is mostly carbon dioxide, and reacts it with a small amount of hydrogen that the spacecraft has brought from Earth. The result is a large supply of methane and oxygen – ideal rocket propellants. Now you have a fully fuelled Earth

The International Space Station heads off into a glorious sunset – one of the most beautiful aspects of spaceflight, according to many of the astronauts.

return vehicle sitting waiting on the surface of Mars.

'Once that's done, you shoot the crew out to Mars in a basic habitation module, like the one on Devon Island. They use it as their house while they explore Mars for a year and a half, and at the end of that time they'll get in the Earth return vehicle and fly back home. They'll leave the habitat behind on Mars, so each time you add another habitat to the base – building up the beginning of the first human settlement on a new world.'

Zubrin's Mars Direct plan comes with a price-tag of only $50 billion, and with further attention to design it could be whittled down to as little as $20 to $30 billion. And while NASA initially viewed the scheme with scepticism, some of the senior officials eventually came on side. It's widely believed that Mars Direct did have a profound influence on NASA's plans to send humans to Mars, although the team in the Exploration Office won't let Zubrin have it all his own way. 'He's more of a creative chef,' suggests John Connolly. 'He didn't really invent the whole idea. He's just added a pinch of salt to a recipe that's already been there.'

Irrespective as to who should claim credit for the architecture of the mission, one thing is certain: the long trek to Mars will require a new approach to living and working in space. Life-support systems, for instance, will need to be completely redesigned. Unlike the 'open-loop' life-support systems used on the Space Station – which rely on the space shuttle and Russian vehicles to bring up water and supplies – the Mars trip will require a closed system, which recycles water and wastes on the long mission. Brenda Ward reveals that the research is quite advanced. 'These aren't just paper studies. There are currently developments under way to take more advanced life-support systems to the Space Station, and even to replace some of the Station's systems.'

Meanwhile, back on the ground in Houston, work on life-support systems continues apace. Don Henninger is Chief Scientist for Advanced Life Support. 'For every person that we have in space, it takes 12,000 kilogrammes per year to supply that person with fresh water, food, clothes and so on. Of that 12,000 kilogrammes, over 10,000 is water. So the technology we can develop to recycle water has tremendous impacts on the overall mass of the mission.'

'For every person that we have in space, it takes 12,000 kilogrammes per year to supply that person with fresh water, food, clothes and so on. Of that 12,000 kilogrammes, over 10,000 is water'

Henninger is currently experimenting with human crews living in atmospherically sealed chambers to see if recycling can work. 'The longest test we've done is ninety-one days, where we've recycled air and water. But we're currently building a new test facilty called Bioplex, which will do more comprehensive recycling – like processing solid waste as well as water and air, and also producing crops for food.'

Bioplex will be big. Consisting of five cylindrical modules, thirty-seven feet long and fifteen feet high, it is about the same size as a Space Station module. A tunnel connecting the modules will allow the crew access to the whole structure, where they will live for periods of over a year.

Timescales like this underline the fact that we need more than just innovative technology. Going to Mars isn't the same as the three-day hop to the Moon and back: it takes the best part of a year to get there. 'It's almost a three-year trip by the time you really work out the numbers,' explains Jeff Hoffman. 'Mars and the Earth only get close together every twenty-six months. So you go there during one of these close encounters, and you come back during the next.'

One way to get around the time problem is to develop a completely new propulsion system – one that makes your craft go a lot faster than the chemical fuels that drive space vehicles today. Former Soviet space supremo Roald Sagdeev and his Greek colleague Konstantin Karavasilis are working on this one, in conjunction with the engineers at Houston. Sagdeev outlines the system. 'You combine two new technologies – nuclear energy and electric propulsion. And then you can compress the time of the flight, and make the initial mass of the spacecraft smaller than it would be with chemical fuel.'

'We're talking about a four-month mission,' adds Karavasilis. 'Although nuclear power is a very politically sensitive issue.'

This is the least of the worries confronting the pioneers who are eager to send humans to Mars. 'I think that there are a lot of different requirements for humans to fly to Mars,' muses Sagdeev. 'Very different from what they experience in orbital stations, because the magnetic field of our planet protects them from cosmic rays. In interplanetary space, nothing can protect them – not even the walls of the spacecraft. Some of the experts evaluating the risk were even comparing the dosage of radiation during a one-year flight with the radiation dosage which the heroic firemen in the Chernobyl accident received.' Don Henninger is working with his water-recycling system to reduce radiation exposure. 'Water's a very good radiation protection device, and it's likely that we would engineer some sort of

> His description of spaceflight makes it sound like something to die for

water tank to provide protection for the crew.' His colleague Doug Cooke agrees. 'Water works very well. You think of shielding being lead or aluminium. The problem with cosmic rays is that they are made of heavy particles – so when they hit these materials, they spin off secondary particles that may be even more damaging than the initial ones.'

'We're very interested in understanding the biological consequences of long-duration radiation exposure,' explains Dave Rhys-Williams, Director of the Space and Life Sciences Division at the Johnson Space Center. 'Then we can develop the appropriate countermeasures to mitigate those effects.'

Canadian-born Rhys-Williams couldn't be more qualified to head up his division. He has the longest list of medical credentials we have ever seen, including time spent as an emergency physician and as a member of the Air Ambulance Committee. Now, he researches into life sciences and space medicine.

Rhys-Williams is also an astronaut. So far, he has flown only once, on a sixteen-day mission aboard space shuttle Columbia in 1998. 'After that experience, I'm very excited about the

Cosmonaut Yuri Romanenko working out on the Mir Space Station in 1987. Vigorous exercise is essential in long-duration spaceflight, otherwise bones weaken dramatically.

opportunity to be able to go back and – in particular – to go to the International Space Station.'

His description of spaceflight makes it sound like something to die for. 'It was absolutely incredible. Being able to look out from the flight deck of the orbiter through the windows, down a hundred and fifty miles to the surface of the Earth – and realising you were there eight minutes ago – it's... quite an experience.

'There are so many different things I could say about being in orbit, but maybe to capture it for you I'll just describe a moment when I was floating on the flight deck listening to Vivaldi's *Four Seasons*. There was a spectacular sunset. On Earth, you can have spectacular sunsets, but they're just in front of you, so to speak. In space, you can see the Sun setting and it casts an array of colours over what looks like a quarter of the planet. As I watched the Sun go down, I thought – here am I, out in space, exploring space – what an incredible honour to have been chosen as part of the team that's being able to do this.

'When it got dark, I started looking at the stars, and I thought – the stars don't look any closer than they do on the ground. And I realised at that moment that – even though we were doing incredible things – how much further we've really got to go, to go on and explore so we can really understand space itself.'

As a doctor and an astronaut, Rhys-Williams is acutely aware of the changes to the body that take place during spaceflight – especially during long-duration stays in space. Some cosmonauts returning to Earth after several months on the Mir space station had to be stretchered off the landing craft because they were so weak. Despite hours of exercising, muscles – unsupported by gravity – become enfeebled in space. And space gets into your bones, literally. Somehow, the microgravity environment accelerates calcium loss, leading to brittle bones and premature osteoporosis. If a three-year human Mars mission is to happen in the near future, using technology not greatly ahead of our own, then how will we cope with the long-term effects of weightlessness?

On the long trip to Mars, with no prospect of an immediate return, how would the crew cope?

'Well, there's a couple of options here,' observes Dave Rhys-Williams. 'One would be to create a spacecraft that has partial gravity at all times, by spinning the vehicle. Another is to have exercise equipment that can create a partial gravitational field. We've got something we're working on right now called the human-powered centrifuge. Imagine cycling round a circular track. As you cycle, you develop a centrifugal force which creates a partial gravitational field – that's because the field at your head is less than that at your feet.'

Of the hundreds of astronauts who have travelled beyond the Earth, dozens have been spacesick – a condition caused by the body's balance systems being confused in zero-gravity – but none has been the victim of a medical emergency. On the long trip to Mars, with no prospect of an immediate return, how would the crew cope with such a dramatic situation – and how would they prepare for it?

'Well, that's a very exciting question,' replies Rhys-Williams, the surgeon in him coming to the fore. 'We need to differentiate the procedures you'd do in zero-g from those you'd do in a partial gravitational environment, and we're trying to learn about both right now. In fact, on Earth you can do a lot of preparation – I was going to say groundwork... You can validate those approaches in the KC-135 [the military transporter plane known to its aficionados as the 'vomit comet'], where you can get very short periods of microgravity. But you really have to go into space to verify those procedures, and that's why we're looking forward to the Space Station.'

Ernest Shackleton's ship *The Endurance*, wedged in ice in Antarctica's Weddell Sea in 1915. Shackleton was a brilliant team-leader, encouraging his crew to row 800 miles in a lifeboat to the island of South Georgia to seek help.

Suppose the unthinkable happened, and a crew member came down with appendicitis? 'Ten to fifteen years down the road, I think we'd be using a computerised-assist device to make the diagnosis, and using a whole series of non-invasive smart sensors to provide the laboratory information to confirm it. We would then have the crew medical officer first practising the surgical procedure in zero-g on a virtual reality surgical simulator, perhaps with holographic guidance from a remotely located surgeon on Earth. And then we would go ahead with the actual procedure in space, using a minimally invasive keyhole surgical approach.'

But even if the crew is in good health, and the spacecraft is working perfectly, one thing cannot be predicted on the long mission to Mars: how the team members are going to adjust to such a new situation psychologically. 'The psychological issues are going to be very, very critical for us,' admits Rhys-Williams. 'There are big medical issues we face in a mission like this. Radiation, of course, is right up there, and the ability to provide autonomous clinical care is right up there. But behavioural human issues are also right up there, and – depending on who you talk to – some people would say they're probably the most significant issue.

But even if the crew is in good health...one thing cannot be predicted...how the team members are going to adjust to such a new situation psychologically

'The difference you would experience on a mission to Mars – different from any other space mission so far – is that you would see the Earth getting smaller, smaller and smaller until it's finally a point. It would be no longer recognisable to the naked eye as the Earth, and that would be a very profound moment for the crew. They would realise that they were experiencing the most extreme isolation that any human could have experienced in the history of human exploration. So I think it's very important for us to learn all about the behavioural issues of exploration and how we can best prepare crew members for them.'

This is an area that Jack Stuster has been investigating for some time. Vice-president and senior scientist at Anacapa Science Incorporated in Santa Barbara, Stuster has studied human performance right across the board – from nuclear power plant technicians, through US Navy teams working under extreme conditions, to NASA astronauts.

'As you head towards Mars, the Earth gets smaller and smaller, and after a while you're alone in the Solar System'

Stuster considers the flight to Mars as being much more like a long sea voyage than a space shuttle flight, or even an Antarctic winter-over. As a result, he has taken a keen interest in the accounts of explorers from Columbus onwards. 'One of the things about isolation and confinement,' he reflects, 'is that several things happen regardless of where the isolation and confinement are – whether in Antarctica, aboard a ship, or in a spacecraft. One is that trivial issues are exaggerated beyond all reasonable proportion. People blow up over some tiny conflict, say over the fax machine or something, and an hour or so later they wonder what the hell went on.

'They usually attribute it to a frailty in themselves, when, in fact, it's a condition imposed on them by the environment, by cumulative stress, and by the unrelenting proximity of one's comrades.'

Nowhere have these kinds of problems been more publicly obvious than on the Mir Space Station. Cooped up in the noisy, ailing spacecraft for months on end, cosmonauts fell out to the extent that they refused to speak to one another. In the later days of visits from American astronauts, the culture differences between the two nations grew into a yawning gulf, leaving the Americans frustrated and alone.

One way around the issue – although it will never go away – is to choose a crew whose members are highly compatible with one another. Stuster is a great admirer of the polar explorers Ernest Shackleton and Fridtjov Nansen in the way they selected their crews. 'Shackleton was expert at establishing compatibilty. He would interview potential members of his expeditions in a manner that was designed to put them off balance. If they reacted in a defensive way, that might indicate they wouldn't be the best team-mates when things got tough. But if they responded with humour, it showed they were the kind of person who could get along when the chips were down.'

The choice of leader for a potential human Mars mission is also crucially important. But – with social mores changing – it wouldn't be a Shackleton or a Nansen. 'There are data to indicate that good leadership is more important than amenities or good habitability,' Stuster points out. 'However, the leader on a Mars mission will not be a leader in the sense of a nineteenth- or early twentieth-century expedition. The leader will be more of the facilitator. Every member of the crew will be a senior professional in their field – it's not quite the same as on previous expeditions. So the leader will have to have very well-developed interpersonal skills, and to lead by consensus. But the leader will have to have complete authority when complete authority is required.'

Stuster believes that the putative crew of the first human Mars mission should train together for long periods of time to build up the team spirit. For the three-year mission, he advocates a previous stay of perhaps four months on the International Space Station, to see how the astronauts adapt.

But even then, some of the most skilled and competent astronauts might prove unsuitable for the long flight to the Red Planet. 'I gave an after-dinner talk on future space expeditions, and a former very senior astronaut came up to me afterwards,' recounts Stuster. 'He said, "You know – I always felt very comfortable on board the Shuttle: I was part of the Earth, Moon, Shuttle system. But I've thought about what it would be like to go to Mars, and I don't know if I'm the right person for the job. As you head towards Mars, the Earth gets smaller and smaller, and after a while you're alone in the Solar System."

'This man was extremely self-aware,' observes Stuster, 'and I hope that others might recognise their own limitations with such a situation.'

Stuster believes that the ideal candidates are people who are not too self-aware, but instead can subsume their personalities to the team effort. They will also have to be tolerant of lack of privacy and personal space. 'Imagine travelling around England in a motor home with five other people for three years and you can't go outside.'

As well as working on the design of private quarters within a Mars-bound spacecraft, Stuster is suggesting other ways in which individual members of the crew can attempt to get away from it all. 'In a recent training session for astronauts preparing for Space Station missions and beyond, I suggested historical novels – Patrick O'Brien, Aubrey Maturin. Twenty books are set in the Napoleonic era, about a

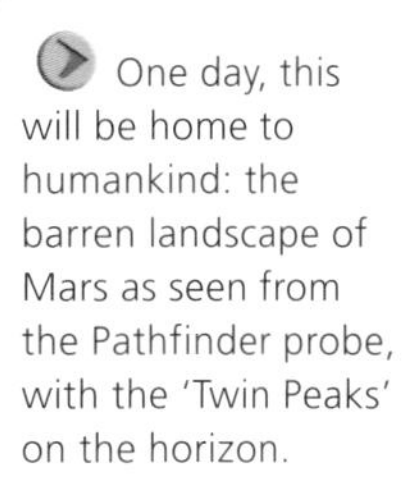

One day, this will be home to humankind: the barren landscape of Mars as seen from the Pathfinder probe, with the 'Twin Peaks' on the horizon.

Royal Navy officer and his friend who's a surgeon. You trace the friendship of these people through adventures and history, and it's the most transporting experience I've ever had.'

Inevitably the crew who travel to Mars will be mixed – which brings up the question of relationships developing on board. 'Romance occurs – it's a natural phenomenon,' admits Stuster. 'I guess it could occur among such a crew. But I hope that the people's motivation and recognition of the bigger issues would make them behave professionally. One way around this is to select husband-and-wife crews, so the whole sex-in-space thing would be over and people can stop speculating about it. I don't know if you saw that recent Disney movie about a mission to Mars, where there was a most touching scene of a husband-and-wife team dancing in the weightless environment of their spacecraft.'

But people will always be human. Last year, a multinational crew undergoing Space Station simulation trials in Moscow asked to be separated from their Russian colleagues because of their behaviour. One of the Russian crew attempted to kiss Judith Lapierre, a French-Canadian physician, despite her resistance.

'I was talking with her recently,' says space communicator Jim Oberg. 'Her problem was not so much the incidents themselves, which were short-lived and relatively self-contained. Her problems were in the reaction of the Russian management to her complaints – which was basically, get used to it, lady.

'Mistakes will happen and they do happen. You can't assume there'll be no interpersonal tensions. You have to have a setup with leadership at all levels that can handle these events when they happen – handle them constructively and then move on. Space is the ultimate judge of reality. You can bluff your way through politics, you can bluff your way through religion, history, literature, even medicine. But you can't bluff mother nature when it comes to spaceflight – people die when other people get into self-delusion.

'Imagine travelling around England in a motor home with five other people for three years and you can't go outside'

'We will keep having these problems, with the test chambers, the harassment, the cover-ups and the disasters. It's like the bumper sticker says: God forgives, man forgives, nature never.'

In the light of all these difficulties of sending humans to Mars – the design of the spacecraft, its life-support systems, the length of the flight, radiation damage to the crew, interpersonal problems on board, and lack of political will to launch in the first place – does Jim Oberg believe that we will ever see people bound for the Red Planet? Or do we need another trigger, like the original space race? 'The space race wasn't simply a matter of competing with someone else. It was a matter of fear of the dominance of another group – of losing your own reputation in the world.

'Now we're racing something else. We're racing ignorance, racing our own cultural future – whether stagnation or reinvigoration. We need to create a climate where it would seem unnatural, in the long run, *not* to go to Mars.' The team at the Exploration Office are following the same agenda. 'We're trying to find out what are the triggers,' admits Hum Mandell. 'We're doing research to find out how we can better communicate our messages and how we can better involve the public in what we do.' And they're working on upping the political will. 'There was a former astronaut who ran for senator in Florida,' explains Jeff George. 'Part of his platform was that he'd push for a human mission to Mars – and now he's a senator-elect.'

'A lot depends on political will,' agrees astronaut Jeff Hoffman. 'But what I'd like to see is for the American administration and hopefully our European, Japanese and worldwide partners to make a commitment at least to start working on the technologies to enable human exploration beyond low Earth-orbit.'

When it happens, the first human Mars mission will almost certainly be an international one – the costs for one nation alone would be too much to bear, and will continue to be until interplanetary travel becomes routine. 'I think the will is going to come out of the creative need to have an international, cooperative project,' says Lou Friedmann of the Planetary Society. 'It will be to accomplish something great – to drive technology and human exploration. It will be like the great engineering projects such as the Suez Canal or the Panama Canal.'

So – when is it going to happen?

'I think humans could get there as early as 2010 if there was a presidential decision and astronomy pushed for it,' asserts Pascal Lee.

Arthur C. Clarke: 'My blind guess is 2020, but of course it's not a technological question; it's entirely a political one.'

'It's very possible we're still a century away from getting people to Mars,' warns Ken Edgett. 'Look, it's been over thirty years since the first men walked on the Moon.'

'I don't believe the person who will be the first human to walk on Mars has been born yet,' echoes Mike Malin.

'It could be as soon as five or ten years if there's a sudden push to go, or it could be as long as forty years if we wait for the economics to become more favourable,' reflects Chris McKay.

'I think, seriously, you're talking about twenty years from now,' says Steve Squyres.

Whenever it does take place, the first human trip to Mars will have no shortage of volunteers. Although geologist Monica Grady would prefer to wait until technology has advanced to the beam-me-up *Star Trek* stage, she would still relish the feeling of breathing in another world. 'It's like, what does it sound like, what does it smell like, what does it feel like, what does it feel like underfoot – I mean, how crunchy is the soil, what's the wind like, does it whistle round the rocky outcrops, even though the atmosphere's so thin? What's the light like when you're so far from the Sun? What colour is the atmosphere with all that

'We need to create a climate where it would seem unnatural, in the long run, *not* to go to Mars'

dust in it? What do the rocks look like? I would find it incredibly stimulating and exciting to stand on an alien planet. But the thought of the journey to get there – no, couldn't do it.'

But astronaut-physician Dave Rhys-Williams looks foward to the first human mission with unconcealed excitement. 'Going to Mars would be absolutely incredible. I think it would be a great honour to be part of a mission like that. I have a shelf in my office which I call the "making-the-impossible-possible shelf", where I have a scanning electron micrograph of the potentially fossilised bacteria that was found in the meteorite in the Antarctic.

'We're going to see whether or not we're alone... And that is such a profound thing for humans to be involved in'

'And I think, can you imagine leaving Earth to go and explore the surface of Mars looking to determine whether or not life may have existed there? And perhaps it sounds like a bold assertion, and maybe it is, but to me that really represents the next phase of human evolution – where we as a human species are reaching out to explore and understand the origins and diversity of life within our Solar System. We're going to see whether or not we're alone, or whether in fact life may have existed elsewhere. And that is such a profound thing for humans to be involved in.'

Even Jack Stuster, who knows more than anyone about the psychological and behavioural problems of sending a crew to Mars, would sign up immediately. 'I'd go in a heartbeat – if I could take my family with me.'

Planetary scientist and field geologist Nathalie Cabrol would also not hesitate to volunteer. And she can also anticipate a time when all the initiatives that we are now setting in train will fundamentally change our perception of ourselves. 'Imagine a day in the future, when a baby is born on Mars for the first time. It's a person who has never seen a sunset on Earth. All they know about Earth is that it's a small blue dot in the sky.

'On that day, an entirely new species will be born.'

SECOND
HOME

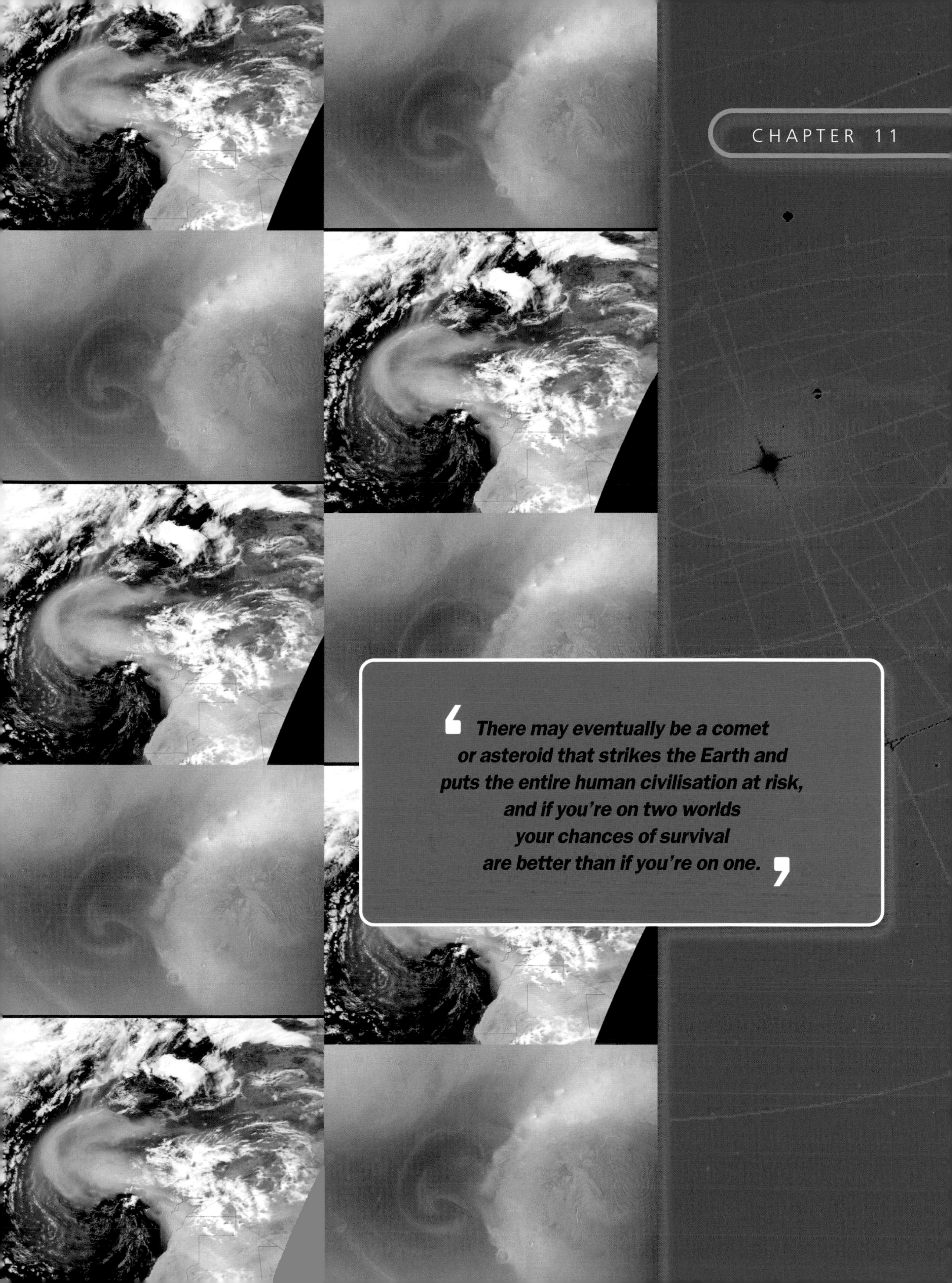
CHAPTER 11
There may eventually be a comet or asteroid that strikes the Earth and puts the entire human civilisation at risk, and if you're on two worlds your chances of survival are better than if you're on one.

AD 5043. Across the sky stretches the tail of a vast comet, the ancient sign of doom. From its haunts way beyond the realm of the planets, a huge block of ice is on collision course with planet Earth.

It's not the first time that the third planet has been in the cosmic firing line. Sixty-five million years ago, a comet smashed into the region we now call Mexico, and the dinosaurs did not live to tell the tale. This time, the human race is the target.

From the Earth, a fleet of rockets rises to meet the cosmic menace. Nuclear warheads blast the comet's icy core. In vain hope do humans wait to see the comet deflected from its path. The nuclear forces shatter the comet nucleus into a thousand pieces – a rifle bullet turned into shotgun pellets, but still with the power to destroy Earth's ecosphere.

A vast fireball erupts from Earth. The comet's impact has instantly charred half of the once-blue planet. Giant tidal waves sweep round the globe, sweeping the rest of civilisation to destruction. Plants and insects will recolonise the Earth, but higher animals and mankind have gone.

By the seas of Mars, humans weep.

'Being on a second world gives you a kind of long-term insurance policy,' says Chris Chyba, an astrobiologist committed to aiding mankind as well as seeking out aliens.

From the Earth, a fleet of rockets rises to meet the cosmic menace. Nuclear warheads blast the comet's icy core

'There may eventually be a comet or asteroid that strikes the Earth and puts the entire human civilisation at risk,' Chyba continues, 'and if you're on two worlds your chances of survival are better than if you're on one.'

Doomsday Earth: fragments of a disrupted comet smash into our planet in this artwork, destroying all higher forms of life – including humans. We will need a second home in order to survive.

Chyba is a keen supporter of human expansion into space, but preferably after we've sorted out the problems of planet Earth. 'In the immediate term, the best way to perpetuate the human species is to make sure we're taking care of our own planet.'

But who's to say that we'll ever defuse the complex and explosive issues of the Earth's political systems? David Wynn-Williams – a British astrobiologist working in the Antarctic – warns: 'There's a possibility that the human species will be stupid enough to virtually annihilate each other by nuclear weapons. In those circumstances, if you want humanity to survive, you'd do everything within your power to do it – including living on Mars.'

And John Rummel, NASA's Planetary Protection Officer, sees a move to Mars as a natural part of human expansion. 'As Earth gets more and more crowded, it'll be more and more desirable to live on other planets. Because – if nothing else – space has got wide-open spaces, and if you want to get away from everybody else it's a great place to go!'

And of all the planets in the Solar System, Mars has by far the most desirable real estate. Mercury and Venus swelter in the Sun's heat; Jupiter and Saturn are giant gassy bodies without any solid surface; and the moons of the outer planets are frozen and wracked with radiation trapped around their parent worlds.

Martian weatherman Rich Zurek, from NASA's Jet Propulsion Laboratory, extols the virtues of the Red Planet. 'Mars is a place you can go: you can stand there, you can look out. It's got an atmosphere; not a very thick atmosphere, but it's thick enough to help protect from small meteoroids. And the atmosphere itself provides a resource. You could take the carbon dioxide, break it into its components, one of which is oxygen. And if we can find water, that would be a tremendous resource. So it's the logical place to go, beyond the Earth–Moon system.'

Lunar missions – like this Apollo 17 expedition in 1972, where geologist Jack Schmitt checks out a huge boulder – can be supported by supplies from Earth. But future Mars explorers will have to 'live off the land'.

When the Apollo astronauts visited the Moon in 1969–72, they took with them everything they needed. Future missions to our Moon can rely, if they must, on supplies and provisions lobbed up on a three-day flight from Earth. But humans on Mars will have no such lifeline. The resources of Mars will be crucial to their survival.

'If every time you went to San Francisco,' says Rummel, 'all the food was shipped from England or Spain, it would be a very expensive place to go. But the fact is they grow the grapes right there, so you don't have to worry about that. It's essential – when we go to Mars – that we get the capability to "live off the land".'

'If you want humanity to survive, you'd do everything within your power to do it – including living on Mars'

To find out how colonists might survive on the red lands of Mars, it's time for a visit to the nerve centre of NASA's plans for human space travel – the Johnson Space Center in Houston, Texas. Here, Doug Cooke is head of the Exploration

Office. 'Yes, it's definitely possible to have self-sustaining colonies on Mars,' he affirms. 'At NASA, our job is to explore and discover. But in the process, we'll be developing hardware and infrastructure that can be used by others.'

Vast iron buildings could tower in Mars's low gravity...[and would] never rust in the planet's oxygen-free atmosphere

In the forefront of this research is Jerry Sanders. He's already cut his teeth on civil engineering plans for the Moon. 'You can take lunar soil,' he says, 'and turn it into a concrete that can be moulded and formed. On the Moon, the big question is: can you do it without water?'

On Mars, however, water shouldn't be a problem: most scientists now expect to find plenty of ice frozen into the planet's soil. In fact, the discovery of 'gullies' on Mars suggests there may be aquifers of liquid water lying not far below the surface.

As well as 'Marscrete', Sanders is looking to other indigenous materials for constructing buildings. The red plains are a planet-wide deposit of iron ore. Vast iron buildings could tower in Mars's low gravity. While iron structures on Earth – like the Golden Gate or Forth Bridges – must be continuously painted to protect them, Mars's buildings will never rust in the planet's oxygen-free atmosphere.

'Nuclear power will eventually be needed on Mars,' explains Sanders. 'You do the initial outpost missions with solar arrays – you just roll out blankets of them. Over time, you'll need something else. Nuclear power provides the backup during a dust storm, and it gives you the potential for growth. You can make nuclear reactors safe to take to Mars – to get them there is not so much a matter of technology as emotion.'

So we have a Mars colony built from concrete and steel, supplied with water from the ground and oxygen created from the air, and powered by nuclear reactors. High above, a network of communications satellites keeps the colonists in touch with the home planet and allows the inhabitants of Earth to play Big Brother with the Mars base. Chris Chyba envisages that 'many of us will have what amounts to kinetic artworks in our living rooms, tuned to the Mars Channel all the time. This will make Mars seem a very familiar place – as the Himalayas are now. And I think it will make us naturally think of ourselves as a species that spans the Solar System.'

The Mars colonists will be largely vegetarian, feeding on crops grown in huge greenhouses on recycled water and nutrients. 'We already have the technology to make green things grow in space, in orbit, on another planetary body,' says Rummel. 'Plants are themselves engineered in such a way that they'll do that without a whole lot of by-your-leave, if we give them the opportunity.'

Sanders utters a note of caution: 'This raises the spectre of contamination of Mars; bring life to Mars and what will be its impact on the ecology?' But Rummel, in his role as Planetary Protection Officer, sees the colonists themselves as a more serious danger. 'People travel with their own microbes,' he says. 'You know, 50,000 cells a day are off-loaded from your body, excluding the alimentary canal. So you've got all this life that travels around with you. If we take that to Mars, it's going to take heroic efforts to prevent contamination.'

While Rummel is concerned about human germs escaping into the Martian environment, most planners are thinking more of the environment's effect on the colonists – in particular, the freezing temperatures, low pressure and deadly ultraviolet radiation from the Sun.

Early science fiction series like *Undersea Kingdom* (1936) predicted that humans would live in vast domed cities under the oceans. In the intervening years there's been little movement in that direction: are cities on Mars equally a pipedream?

Huge cities flourish under glass domes in this artwork of Mars's future. The colonists have started to terraform the surrounding desert: when the atmosphere is thick enough, they will tear down the domes and live in the open air.

'You'll be living in domes that are pressurised,' predicts Zurek, 'so in some ways it will be like living in a space station. But it'll be more familiar: there's a window and an atmosphere you can look out on, and weather that comes and goes, just as there is on Earth. And you'll be able to get out and explore – in a spacesuit – and that sense of exploration is very important.'

Astronaut Jeff Hoffman has experienced living in space. 'The idea of having scientific research stations on Mars permanently occupied is, I think, a realistic goal. The urge that human beings have to explore and live in new environments will take large numbers of people to Mars, if it turns out to be a planet with enough resources to support life.'

But Lou Friedman, president of the Planetary Society – a public pressure group pushing the human exploration of space – sounds a note of caution. 'There's a lot we need to do on Mars, setting up a robotic outpost and then a human outpost and finding out how we adapt to that planet.

'The example I always give,' Friedman continues, 'is pick up a popular science magazine from the 1950s or 1960s, and you see the idea of underwater cities being developed because the Earth was getting too crowded – and you know we don't yet have a permanent human facility under water.'

Hoffman insists it's a matter of timescales. 'Some day it will happen that we see huge cities on Mars,' he predicts, 'but we may be talking a thousand, several thousand years in the future.'

Some people aren't prepared to wait that long. At NASA's Astrobiology Institute, 'Mr Mars' Chris McKay wants to see humans living on Mars within a century. And when McKay says 'living on Mars', he doesn't mean dwelling within concrete bunkers. He's talking about humans inhabiting the planet's wide-open spaces.

'There've been many suggestions over the years for how to warm up Mars and restore it to habitable conditions,' says McKay. 'But most of these were fantastic ideas – like building mirrors the size of Texas.'

Instead, McKay wants to pump Mars's atmosphere full of the kind of gases that create the notorious greenhouse effect on Earth. As the planet warms up from its current average temperature of –55°C, carbon dioxide frozen into its poles will evaporate and thicken up the atmosphere. Eventually, the planet will become a second Earth.

The idea of 'terraforming' Mars ironically came originally from a British scientist well known for his warnings about tampering with planet Earth – Gaia-inventor Jim Lovelock

The idea of 'terraforming' Mars ironically came originally from a British scientist well known for his warnings about tampering with planet Earth – Gaia-inventor Jim Lovelock.

The terraforming dream envisages a warm atmosphere cocooning the Red Planet. With Mars unfrozen, streams and rivers will flow; lakes and seas will fill

'I wrote a book with Mike Allaby on it in 1984, called *The Greening of Mars*,' says Lovelock. 'Our view was that an entrepreneur might make money by getting rid of stocks of fluorocarbons and sending them to Mars, where they'd act as a temporary greenhouse and perhaps lift off water and carbon dioxide.'

Since then, McKay has taken up the terraforming baton. In the autumn of 2000 he held an international conference at the Astrobiology Institute's home base, the Ames Research Center. 'The major result to come out of the conference,' McKay sums up, 'was a closer look at the use of greenhouse gases to warm Mars. It's really plausible that they could warm up Mars on a timescale of fifty to a hundred years. That's a pretty interesting and optimistic result.'

McKay's colleagues have been investigating 'super-greenhouse gases'. The stuff of nightmares for environmentalists on Earth, these substances can heat up a planet by 50°C in less than a century.

At the Massachusetts Institute of Technology, outside Boston, undergraduate Margarita Marinova is driving forward McKay's vision. She has already identified chemicals that will do the job of trapping heat in Mars's atmosphere, and worked out how to make these perfluorocarbons from the minerals present on the Red Planet. 'If you had a hundred factories,' she calculates, 'each having the energy of a nuclear reactor, you could warm Mars by six to eight degrees in a hundred years. With more efficient artificial super-greenhouse gases, perhaps the timescale could be cut to a decade.'

So far, these ultimate super-greenhouse gases have yet to be made; they exist only in a computer. But – to give a taste of McKay's future Mars – conference delegate Mimi Gerstell announced that just a whiff of the gases she's 'invented' would trap almost all of the Sun's heat falling on the planet. 'The chemicals we looked at are sci-fi right now,' she says, 'but chemists say they should be easy to make.'

The terraforming dream envisages a warm atmosphere cocooning the Red Planet. With Mars unfrozen, streams and rivers will flow; lakes and seas will fill. In this future Eden, humans are already a fixture. Now comes the time to populate this world with plants and living creatures of every kind.

Phobos orbits over a Mars that's being transformed into a wetter and bluer world.

An unspoilt and pristine environment – a cosmic Antarctica – Mars hangs in front of the Milky Way in this artist's impression. Do we have any right to inflict on this world the same environmental damage we've already wreaked on Earth?

With palm trees growing beside the seas of Mars, and cicadas buzzing away the last of daylight, mankind would then have a second home in the cosmos – a safe retreat if Earth itself were threatened

Antarctic explorer Charlie Cockell believes that insects could be introduced early on, while Mars's atmosphere is still building up. In his Cambridge lab, he's subjected insects to lower and lower pressures. 'Many behave quite normally at only 20 per cent of Earth's atmosphere pressures,' Cockell reports, 'while milkweed bugs mate and cockroaches lay eggs at only 10 per cent pressure.'

Along with insects, in McKay's vision, will come trees. With their great expanses of leaves, trees can best mop up the available sunlight and begin to transform Mars's carbon dioxide atmosphere into breathable oxygen. With palm trees growing beside the seas of Mars, and cicadas buzzing away the last of daylight, mankind would then have a second home in the cosmos – a safe retreat if Earth itself were threatened.

But not everyone agrees that we'll be able to replay Genesis so easily. Jeff Hoffman, who's been privileged to view the green Earth from space, is not entirely convinced. 'Until we really understand the planetary processes which caused Mars and the Earth to take such different histories, I'm a little bit sceptical of any conclusions about the possibility of terraforming.'

And Cockell, despite his ideas on introducing insects to Mars, also believes we are a long way from understanding how to terraform the planet. 'Terraforming is essentially something that enthuses engineers,' he explains, 'who tend to have this view that just mix something in a test tube and – hey, presto – instant biosphere. From my own research I really don't think that what they want to do on Mars they're actually going to end up with.'

Referring to the great science fiction trilogy on terraforming, *Red Mars*, *Green Mars* and *Blue Mars*, by American writer Kim Stanley Robinson, Cockell continues: 'I've often proposed writing a book called *Black Mars*, to be the follow-on to this series. It would describe what happens when the whole terraforming exercise goes completely wrong and they all end up with this horrible planet where there are viruses and bacteria doing everything they're not supposed to do!'

Unlike the engineers that Cockell cites, we find little enthusiasm for terraforming among the NASA planners who are looking into a future human expedition to Mars. 'There's some scientific validity to the idea,' says Hum Mandell, 'but there's nothing we could do practically with any energy sources that we currently know about within our lifetimes.'

'Terraforming is something that's going to happen over a very large timescale,' adds John Connolly, 'a geological kind of timescale, not human timescales. I mean, you could argue that for the entirety that humans have been on this planet we've made an almost imperceptible change in the climate and the terrain – and that's so many million of us working our hardest!'

Geologist Pascal Lee agrees. A veteran of Devon Island in the Arctic, the nearest environment to Mars on Earth, Lee says: 'I shy away from spending too much time thinking about terraforming, because to me that's the challenge of another time, another age.'

The difference of opinion of whether we can terraform Mars, however, pales before the passionate arguments over whether we should even try. At one end of the spectrum is Jim Lovelock, original exponent of the idea in his book *The Greening of Mars*. Surprisingly, he declares: 'I think it's obscene. The whole idea is almost wicked to me that we've overcrowded and messed up the Earth to a point that we should look at Mars as the place we next go to. We should be turning round and seeing what we could do to make the Earth a better place to live.'

Arthur C. Clarke has also written a book on terraforming Mars – *The Snows of Olympus.* He succinctly echoes Lovelock's opinion: 'Terraform Mars? Well, I hope we've terraformed Earth first.'

Far future 1. A space shuttle docks in the low gravity of Phobos base to disembark passengers from a terraformed Mars. In a massive exercise in planetary engineering, vegetation is covering most of the once-red planet.

'I really don't like the idea of terraforming Mars – I think we should keep some of the things we have in our Solar System pristine,' says Everett Gibson from NASA's centre in Houston. His colleague Kathie Thomas-Keprta adds: 'Look at some of the pristine places that we still have on Earth, where we're trying to keep people out and preserve our natural environment. I feel the same about Mars – I'd like to preserve it in the state it's in now.'

Antarctic explorer David Wynn-Williams has seen more of Earth's last remaining wilderness than most. 'Ethically at the moment I'd say terraforming Mars is wrong,' he judges, 'just like colonising the whole of Antarctica is wrong. Antarctica is a fantastic place for science; Mars is a fantastic place for science. We need to find out things like the origin of life, the limits of life, all sorts of fundamental questions, which we could answer by studying Mars before we spoil it.'

'What are the rights of those Martians, even if they are only microbes?'

For most scientists, in fact, the question of life on Mars comes before any opinion on terraforming. Chris McKay has pondered the ethical issues long and hard. 'Think back to the time when Mars had a habitable biosphere,' he argues. 'If Mars really had life then, maybe the bugs frozen in the soil can come back to life and we'd have Mars for the Martians.

'If it doesn't have any life of its own, then I think it would be appropriate for Earth to share its life with Mars. So I see it really as a restoration rather than a creation of something where there was nothing before.'

Jill Tarter, as head of Project Phoenix – which is looking out for radio signals from distant aliens – has equally strong opinions on the ethics of contact. 'We need to find out whether there are any Martians there first. I'm one of those folks who actually think we should ask the question: what are the rights of those Martians, even if they are only microbes? Should we be imposing our biology on another planet that is already biologically active?' Or, as Planetary Protection Officer John Rummel quips: 'Green slime has rights as well!'

In the colonisation of Mars, the whole ethical issue of humans versus Martians is likely to come up well before any full-blown terraforming is under way. It's a problem as basic as plumbing.

The first Mars colonists will have to drill wells for drinking-water – yet subterranean water is the most likely place in which to find living Martian bugs. A terrestrial water engineer would take the utmost pains to sterilise any water supply that may contain microbes.

NASA engineer Jerry Sanders has had to confront this problem head-on. 'At a meeting I jokingly said, "OK, you've drilled, you've found water, you've found life – now can I run it through an electrolyser and kill it?"' Sanders recalls. 'And at first, everybody laughed, and then they suddenly got quiet and thought seriously about it. Somebody said, "Well, until you find the next pool of water that has life – the same life – in it, you can't touch the first one."'

If microbes are widespread on the Red Planet, then astronaut Jeff Hoffman thinks incoming humans will be able to come to an accommodation with the native wildlife. 'Perhaps you could establish nature parks on Mars while you develop human civilisation in other parts.'

Life on Mars expert Chris Chyba agrees. 'If there is Martian life, then we have to respect that biosphere. I don't think that rules out human exploration or human settlements, but it does say there's a lot we need to understand better before we start building there, certainly before we start on any sort of terraforming project.'

Colin Pillinger, who heads up Britain's Beagle 2 mission to Mars, takes a stronger stance. 'I just think people wouldn't do it. There'd be an enormous ethical backlash if you changed the environment of Mars and tried to convert it to an Earth-like planet, when it was the only example we have of another form of life. It's incredible anybody would even contemplate doing it.'

Yet it's not too hard to find people, both in NASA and outside, who uninhibitedly espouse colonisation and terraforming. Robert Zubrin is the leader of the Mars Society, a pressure group for exploring and settling the Red Planet.

'The nature of life is to take barren environments and transform them into those that are firmly for the development and propagation of life,' he begins. 'No sooner does any barren place appear on Earth than life colonises it. Take Hawaii, coming out of the Pacific Ocean. Birds drop seeds, plants grow, Polynesians show up, Europeans appear – and build hotels.

'We exist on the basis of the gifts given to us by the pioneering spirit of life in the past,' Zubrin continues, 'and in a certain sense I believe we owe it to life for the future to create the same sorts of possibilities for it. So I think that we will terraform Mars. We'd be less than human – we'd be less than life – if we didn't.'

Maggie Zubrin, his wife and partner in running the Mars Society, is equally keen. 'I feel very excited about terraforming,' she enthuses. 'I think it would be a blessing for humans to give such an atmosphere to Mars, where life could thrive. Those little Martian microbes – if they're still existing – could develop their potential as well.'

Her husband is optimistic there'll be little conflict between the terraforming humans and any indigenous bugs. Robert Zubrin cites the case of very ancient microbes living miles beneath the surface of the Earth, and carrying on their aeons-old way of life even though the atmosphere above has changed radically. 'If there is life on Mars today,' he says, 'it's unquestionably in the form of subsurface microbes. If we did terraform the surface of Mars, frankly that would have very little impact on the subsurface habitat.'

'This is a crowded planet...and we are in survival mode – reaching out for new environments'

And at NASA's Astrobiology Institute, French-born geologist Nathalie Cabrol puts Martian

Far future 2. Mars looks no different from this 1970s Viking image – but humans are populating the planet. Our descendants have been genetically engineered to live in the planet's thin air and freezing temperatures…

But the science of genetics may be the key to the survival of humans on Mars

microbes well in second place, compared with the demands of human society. 'This is a crowded planet,' she expostulates, 'and we are in survival mode – reaching out for new environments. This is a call that is inside of us, has always been inside of us, since the day we were created.'

And if our need to settle the Red Planet conflicts with the 'rights' of indigenous Martian bugs? 'Life is not ethical – life is a process which, for survival, kills and conquers. I'm not saying I'm not an ethical person: I'm not going to kill my neighbours. But some people are saying we must go to Mars, without seeing the consequences of this exploration. Living has dramatic consequences.'

Cabrol militates against purely commercial exploitation of Mars, like plans to mine the Red Planet. But she is all in favour of human intervention if it's 'to put an atmosphere there, to give the human species an extension – and not just the human species. They'll probably come along at some point with dogs, with cats, and that would give their evolution an interesting new turn.'

And it may be accelerated evolution, rather than terraforming, that shapes our own future destiny on Mars. The first generations to be born on the Red Planet will live in low gravity. Their bones may be weaker and their muscles less developed. NASA's mission planners in Houston already concede that children of the first colonists may never be able to return to the strong gravity of planet Earth.

Perhaps over the generations, humans on Mars will evolve away from their cousins at home, towards a creature more suited to low gravity, low temperatures and low pressures. Cabrol sees a future existence on Mars 'as a way for life to evolve, to diversify – as a transition towards something else'.

Her colleague Pascal Lee elaborates: 'It's often said that you could change the planet, and that's at the heart of terraforming – to allow humans as we know them to live there and thrive. But at the same time it's becoming clear that what might catch up with us faster than the engineering capability for terraforming is progress in genetics.'

With the human genome – the 'book of life' – now open for any scientist to read, huge advances are already promised in medical science, from preventing inherited diseases to curing cancers. This ultimate codebook, however, promises much more: an option for changing the nature of humans.

On Earth, genetic manipulation is a double-edged weapon. Dark memories of 1930s eugenics – the totalitarian idea of creating 'better' people – hang over the prospect. But the science of genetics may be the key to the survival of humans on Mars.

'With this progress in genetics,' says Lee, 'we might actually be able to adapt human beings and genetically engineer them to live on Mars – without really changing the planet much.'

Terraforming or genetic engineering? It's too early yet to discern clearly the future course of humans on the Red Planet. But the crystal ball is now beginning to suggest a rather unexpected future. Instead of humans shaping Mars to become a second Earth, Mars may shape humans.

If life on Earth is extinguished in millennia to come, then a creature with long lanky limbs, a thick ultraviolet-protective hide and a huge ribcage enclosing extended lungs – *Homo sapiens martialis* – may become the ultimate heir to the Galaxy.

WORLDS
BEYOND

CHAPTER 12

“The other thing that I remember was so stunning about Mars was being able to see the white polar caps. If you squinted and looked in the eyepiece ever so carefully, you could see polar caps on another planet – which rendered it kin to our own Earth.”

Surrounded by waterfalls, we wait in the huge atrium of the Hyatt Regency Hotel at San Francisco International Airport. Chandeliers shaped like stags' antlers light the enormous wood-panelled space. People come and go from a conference. A sign on the Knuckles Historical Sports Bar boasts that it has been receiving awards since 1992. If you want to find contemporary America, this is the place to be.

One of the twin Keck telescopes (dome on right). Their enormous mirrors are essential tools for flushing out planets around other stars.

It's also the place to be if you want to meet the world's leading planet-hunter. Since 1995, Geoff Marcy and his collaborators have swept the board in discovering planets around other stars. Marcy has just flown in from Hawaii, where he has been using the world's biggest telescope – the Keck – to winkle out more. We arrange to meet up at the best of the airport's hotels to find out how he has got on.

The aptly named Whirpool spiral galaxy, imaged by the Hubble Space Telescope, is a dead-ringer for our Milky Way. How many of its billions of stars have encircling planets?

> **The influence of Mars, as an icon for a living world, goes way beyond our Solar System**

Marcy arrives with a big grin on his face. He's clearly had a good observing run. We ask him to remind us just how many extrasolar planets he has found. 'Well – we've found twenty-seven, and we're about to announce three more. We haven't told anybody, so nobody knows about this yet. We'll announce them at the International Astronomical Union meeting in Manchester next month [August 2000]. We wouldn't normally do this, but they've asked us if we have any big announcements to make, and we said, well... because we don't want too much hullabaloo. But there will be three more.'

As we write, the total number of planets orbiting other stars stands at fifty-three. And thanks to the never-ending fascination with life on Mars, this translates to fifty-three possible places where life might exist. The influence of Mars, as an icon for a living world, goes way beyond our Solar System.

It was Mars that got Geoff Marcy hooked on astronomy. 'When I was fourteen, my parents bought me an old, used telescope – a reflector with a four-inch-diameter mirror. I still have it today – it's my only telescope, actually. And I can remember very vividly crawling out on to the roof of my house at night, and spending the whole night looking at Mars.

'And I remember two things about Mars that were shocking to me, really. In a small telescope you can see this thing that's round – it clearly is a sphere – floating in the darkness of the cosmos. To see the nearest twin to Earth looking so vividly like a little bobbing Christmas ornament in space is just unforgettable.

'The other thing that I remember was so stunning about Mars was being able to see the white polar caps. If you squinted and looked in the eyepiece ever so carefully, you could see polar caps on another planet – which rendered it kin to our own Earth.'

At the same time that the teenage Geoff Marcy was gazing in wonderment at Mars, a few, more senior, astronomers were searching for planetary systems beyond our own. The rationale behind the search was impeccable. The Sun is an average star – one of perhaps 200,000 million in our Galaxy alone. When stars are born out of the huge gas and dust clouds that drift among the wide-open spaces of our Milky Way, encircling planets form as a by-product. So planets are likely to be even more common than stars.

The problem is that stars are big and bright, while planets are small and dark. It's a situation akin to detecting moths around a searchlight when you're stationed in London, and the searchlight is in New York. No way can astronomers see extrasolar planets directly, even with current instrumentation. But, as Geoff Marcy explains: 'The technique is to watch the star carefully. Then see if it wobbles due to the gravitational pull from its planets.'

It sounds easy. But the handful of astronomers who tried to seek out extrasolar planets in the days before today's precision computer-controlled equipment had to make measurements that were up against the very limits. The pull of even a massive planet like Jupiter on its parent star produces only the tiniest of wobbles. Marcy's predecessors, like Peter van de Kamp of the Sproul Observatory in Allegheny, were having to measure displacements of less than one-thousandth of a millimetre on old-fashioned photographic plates. No wonder the results were uncertain – and controversial.

Van de Kamp's best candidate for a star with planets was a nearby 'red dwarf' called Barnard's Star. A dedicated astronomer of the old school, he monitored Barnard's Star for fifty years, and came to the conclusion that it was orbited by two planets roughly the mass of Jupiter and Saturn. The problem was that no other astronomers could duplicate his results. Sadly, it now appears that the 'wobble' seems to have been in his telescope, and not in the star: what he had been measuring so conscientiously for half a century were instrumental defects.

To add insult to injury, hunting for extrasolar planets was hardly considered respectable by the wider astronomical community. 'In the early 1980s, the idea of searching for planets around stars was akin to looking for little green men with your satellite dish,' recalls Geoff Marcy. 'It was like pyramid power. People frankly laughed at the idea – they thought it was like trying to find UFOs and extraterrestrials.'

Little considerations like this did not daunt Marcy from starting his own planet search. In the early 1980s his own career was getting nowhere fast, and he needed a project that would totally absorb him. 'I was twenty-eight, and doing research on the magnetic fields of stars. Frankly, it was not all that grippingly exciting, and I remember being more than a little depressed about the state of affairs. My career was not even exciting to me – so how could I expect it to be exciting to my colleagues or anybody else?

Dusty curtains envelop young stars in the constellation of the Chameleon. This 'interstellar soot' not only forms planets, but it is also rich in the raw materials of life.

'I needed a project that gripped me on a very personal level'

'I remember one day – as a post-doc in Pasadena – taking a shower in my apartment, thinking what was I going to do. And I realised that I was in a sense going downhill – I needed a project that gripped me on a very personal, sort of almost childlike level. I thought to myself, I should try to find a research project that addresses a question I had asked myself as a child. And the question that popped into my mind, with the shower still running, was whether or not there were other planets orbiting the stars we see at night. That's how it all got started.'

Marcy knew that, in order to be successful, he would have to push the 'wobble' technique to its limits – using modern technology. 'My collaborator Paul Butler and I started developing our technique even before we began taking data at the telescope. And – frankly – for eight years we did nothing other than work on our technique. We didn't announce any planets, we didn't find any planets. Instead, we worked on the hardware, the optics, the computer programmes, new telescope optics – and most especially some newfangled algorithms to analyse the light from a star that allowed us to detect the wobble more precisely than other people could.'

'Other people' unfortunately had the same idea. On 6 October 1995 Michel Mayor and

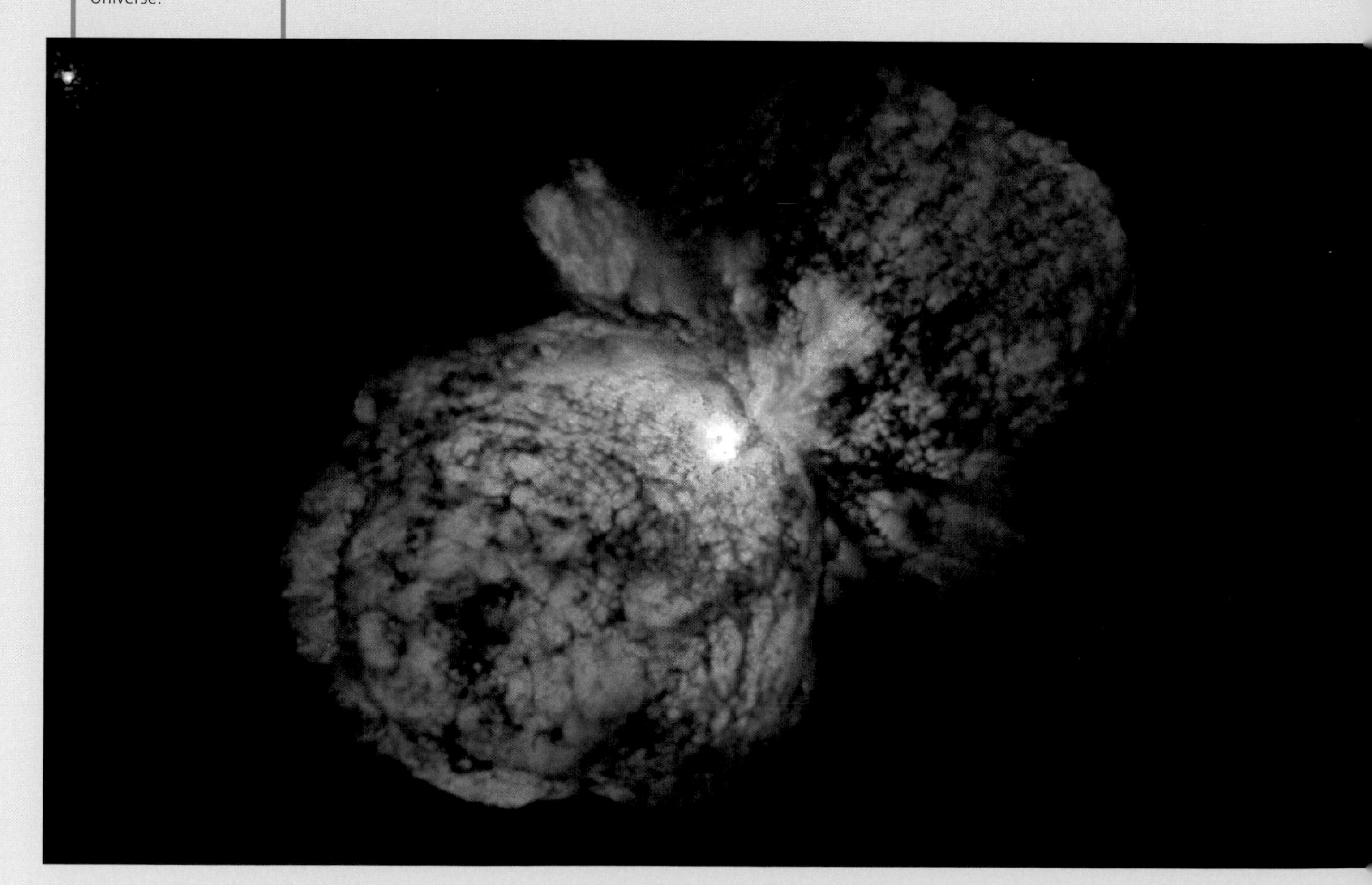

The unstable, supermassive star eta Carinae has already ejected vast amounts of gas. When it eventually explodes as a supernova, heat and pressure will forge exotic elements like gold and silver – the only source of precious metals in the Universe.

Didier Queloz of the Geneva Observatory announced that they had discovered a planet orbiting the sun-like star 51 Pegasi. But Marcy and Butler were not far behind. Three months later, they announced two new planets orbiting 47 Ursae Majoris and 70 Virginis, respectively. Since then, the American team – who are great pals with the Swiss astronomers – have caught up with their European rivals, and have several more suspects in their sights. 'We're monitoring 1,100 stars right now for planets, and some of them show preliminary hints of wobbles that we know will mature. So we watch these "spectroscopic eggs" incubate and wait for them to hatch into planets and we can see the cracks on quite a number of them already.'

Marcy's and Butler's incredible success record buys them time on the world's largest telescope, the Keck on Hawaii. Its vast thirty-three-foot-diameter mirror has the light-collecting area equivalent to half a tennis court – essential for such high-precision measurements. 'It's always a fantastic feeling to use the world's largest telescope,' enthuses Geoff Marcy. 'It's a privilege, and I always feel a little embarrassed that somehow I find myself in a position to use this marvellous instrument that's the product of hundreds of years of technology. And the last observing run was particularly exciting.'

Knowing how astronomers keep themselves awake on their through-the-night vigils, we ask Marcy which audiotapes he was playing in the Keck dome. 'The *Symphonie Fantastique* by Berlioz, plus *Harold in Italy*. Then there was Dvořák's String Quartet and his Piano Quintet.'

How do Marcy and Butler know which stars to target for their planet searches? 'Choosing the stars is tricky,' Marcy admits. 'Many of my colleagues don't realise how much work we put into this. First, we choose stars that are somewhat like the Sun, maybe a little bigger or a little smaller. And we choose stars that are close to our Sun, because they offer us opportunities to follow up our detections with other techniques. For example, if the Hubble Space Telescope is revamped one day, maybe with an occulting bar to block out the light from a star, you might be able to see the reflective light from its planet – but only if the star is close enough to us.'

'It's always a fantastic feeling to use the world's largest telescope'

In addition, Marcy and Butler have a secret weapon – an English undergraduate studying at Sussex University called Kevin Apps. Apps, as a first-year student, had the temerity to e-mail Marcy and ask for a list of his target stars. Apps was fascinated by the work of Marcy's team, and also aware of a new database of stars that had recently been gathered by a revolutionary satellite called Hipparcos. 'I happened to come across a copy of the Hipparcos Catalogue,' explains Apps. 'I checked on the three hundred stars that Marcy and Butler were observing from this catalogue, and found out that thirty or so were unsuitable candidates for finding planets around. Some of them were further away than astronomers had thought, while others are close double stars, which aren't suitable for planets.'

Marcy still reels with embarrassment at the memory. 'When Kevin Apps had the audacity to suggest that thirty or so of our stars were inappropriate, it was like saying that we didn't know how to choose stars – we professional astronomers, at the world's biggest telescopes, who've been planet-hunting for years.'

Mission of the future: floating free of the penetrator that has broken through the thick ice sheet, a hydrobot explores a black steamer on Europa's ocean floor.

But they were soon to acknowledge that Apps was absolutely right. 'We're working daily with Kevin. We couldn't be closer if we were even in the same office. We'd communicate no more frequently than we do via e-mail, and he's – what is it? – 7,000 miles away.

'What's truly phenomenal about Kevin Apps – and you know, as a scientist I hesitate to sound hyperbolic – is that he has the most encyclopedic knowledge of the properties of thousands of stars memorised in his head. Their chemical compositions, their temperatures, their masses, their radii and so on. I don't know how he does it. But he's making a unique contribution to our planet search. He chooses all the stars that we now point our telescopes at, and then once we've found planets he associates the properties of the star with these planets, so that we can see patterns emerging.'

The planets that Marcy and his team are finding are nothing like the Earth or Mars. Even with the team's precision techniques, they cannot detect planets much lower in mass than Jupiter or Saturn – anything more lightweight cannot exert sufficient pull on its parent star to make it wobble. We know that gas giants can't harbour life – so are Marcy's planets stillborn? 'The gas giant planets we're finding – the Jupiters and Saturns – certainly don't offer the opportunity for life,' reflects Marcy. 'But there are two things that are possible, and this is why I can only hum and haw and wave my hands in the air: the Jupiters

The planets that Marcy and his team are finding are nothing like the Earth or Mars

we're finding are obviously the biggest planets in any planetary system, and they may be accompanied by smaller rocky planets. So one possibility is that the Jupiters serve as the signpost of a fully fledged planetary system, the other members of which are currently undetectable.

The mighty planet Jupiter with two of its 28 moons, Io (orange) and Europa (white). Astronomers believe that the moons of Jupiter-like planets orbiting other stars might be fertile hunting-grounds for life.

'I still think the highest chance of finding life in our Solar System would be on Europa'

'The other interesting possibility is that there may well be moons orbiting our Jupiters. And we've learned just in the last five years that the moons orbiting our own Jupiter – in particular Europa and maybe Callisto – could harbour life. It's possible that life in the Milky Way Galaxy is more abundant on the moons of giant planets than on the heart-warming Earth-like planets that we had been thinking would be the only place in which life could form'.

The icy surface of Europa is a magnet for those who are driven to search for life in the Universe. Scientist-diver Oliver Botta of the Scripps Institution of Oceanography is fascinated. 'There's a very strong indication that there is a liquid ocean under the ice surface. One question right now is: how thick is the ice sheet? There are opinions between ten and a hundred kilometres – it's really important that we find out.

'The other question about Europa is: is there enough energy to support a Europan habitat at all? It's further away from the Sun, and so the energy input into Europa is much less than on Earth or on Mars. Here also the opinions are split. Some people say, "Yeah, it's enough", while others say, "No, it's too cold, and there's not enough energy". But I still think the highest chance of finding life in our Solar System would be on Europa.'

'I agree,' enthuses his colleague Danny Glavin. 'I think we're going to have to develop some good techniques to be able to penetrate the surface of Europa and investigate exactly what's in this ocean. And we have a good analogue here on Earth in Lake Vostok, which has been isolated for millions of years from the surface underneath the Antarctic ice sheet. The Russians have actually been drilling down, and they recently stopped a hundred metres above the lake itself, for fear of contamination. I'd sure love to analyse a sample of that lake for amino acids, to see what we could possibly expect on Europa.'

Frank Drake of the SETI Institute in Silicon Valley reveals that he, too, is a scuba diver. 'I think the ocean of Europa's a fascinating place and has the potential for having more complicated life forms than on Mars. I think all the life forms on Mars will be extremely primitive single-celled organisms, but on Europa there's been an opportunity for evolution to occur in the ocean. If there's life there, you might find some very interesting creatures.'

Another Europa fan is NASA's John Rummel. 'It's a fascinating world. Very challenging conditions on the surface, but centimetres below it's an environment where we know Earth-life could be preserved. Conceivably, kilometres below that there's a nice warm liquid water ocean with all sorts of good stuff to eat.'

Rummel is not too perturbed about Europa's energy problem. 'The other three Galilean satellites and Jupiter all interact with Europa, flexing it like it was a piece of clay. As it gets flexed, it gets hot – typical friction. There might be a lot of heat coming out of Europa. There could be hydrothermal vents and mid-ocean ridges for all we know.'

A former naval officer, Rummel muses: 'A nuclear submarine today could probably exist quite nicely in the Europan environment. It's got all the basics – heat and water – and if you have nuclear power, you're pretty much there. But I think a remotely operated vehicle's the first line

'If there are 4.5-billion-year-old briny oceans there...there's likely to be life'

of defence there. It's a long trip time, and there's high radiation at the surface.'

Earth's first calling-card to mysterious Europa will be an orbiting spacecraft scheduled to visit sometime within the next few years. 'The launch for the Europa Orbiter fluctuates anywhere between 2003 and 2007,' explains Rummel. 'When we can afford it, we'll do it. But there are lots of us who would like to send it tomorrow.'

Jill Tarter of the SETI Institute is also confident that Europa and possibly Callisto are worth investigating in depth. 'If there are 4.5-billion-year-old briny oceans there,' she reflects, 'there's likely to be life.'

Jill Tarter has a vested interest in finding life in the Universe. She is Principal Scientist at the SETI Institute – a privately funded institution that concentrates on the search for extraterrestrial life – and has assumed Frank Drake's mantle as the world's leading SETI researcher. Tarter is passionate, compassionate, eloquent and practical. 'What's so terrific is that we suddenly have the technology that allows scientists and engineers to answer the question as to whether there's life in the Universe. For me, I can't imagine doing anything more important.'

The lead character in Carl Sagan's novel *Contact* (which was later filmed, starring Jodie Foster) is widely reputed to be based on Jill Tarter. The story tells of a woman dedicated to finding the first intelligent life forms beyond Earth, and – in parallel – it follows her spiritual search for her much-adored father, who died when she was twelve. She ends up finding them both, although the rite of passage involves travelling through a wormhole.

Tarter is a radio astronomer who, like many others, has followed in Frank Drake's footsteps. In the late 1950s – when radio astronomy was in its infancy – Drake wondered if his radio dish could pick up *artificial* signals from space. After all, he figured, we communicate using radio waves on Earth, because radio waves are cheap and very fast. So why not in space?

Over the years, Drake gathered a group of like-minded scientists around himself. It was slow progress at first, because searching for alien signals was not deemed to be respectable. 'Back in the 1960s, not only the idea of life in the Universe – but also even ideas about the nature of the planets – were considered not very reputable subjects in science,' recalls Drake.

The group – and others who were also inspired by Drake's vision – began to construct highly sensitive receivers that were capable of

The 1,000-foot-diameter Arecibo Telescope in Puerto Rico – the largest in the world – was the venue for NASA's first official search for alien radio signals in 1992.

tuning in to millions of 'extraterrestrial radio stations' simultaneously. At the same time, the group started developing powerful pattern-recognition software that would reveal the alien signal buried in the cosmic noise – akin to finding a needle in a haystack.

Along the way, there were several dramatic false alarms. One, which took place in 1977, was so powerful that an astronomer wrote 'Wow!!' in the margin of the computer printout next to it. The Wow!! signal never reappeared. It was almost certainly interference – probably military. And it was the same for all the other signals.

But by 1992 Drake's group had managed to persuade NASA to give them a modest grant – just $100 million over a ten-year period – to develop their SETI researches. That was the year when we first caught up with Drake's team, at NASA-Ames in Silicon Valley. We were making a TV programme about SETI and, before we got down to filming, did a 'recce' to check out people and locations.

We were fascinated by Drake's team. They had more perspectives on science and humanity than any other scientists we had ever met. Our conversations grew to be very deep and very long – much to the despair of the public affairs unit staff, who have to escort 'foreign nationals' around the campus. One staff member desperately tried to hurry us along. 'Hands off,' warned one of the SETI researchers. 'This isn't your usual TV crew. These are little kiddies running around with PhDs.'

Later that year, the team hooked up their receiver – so big that it had to be housed in a container vehicle – to the largest radio telescope in the world. We were at the opening ceremony of the world's first official SETI initiative at Arecibo, Puerto Rico, on 12 October – Columbus Day. Jill Tarter flung the switch with the words, 'Let the search commence'. The 1,000-foot-diameter

The 1,000-foot-diameter Arecibo radio telescope began listening in to signals from the stars

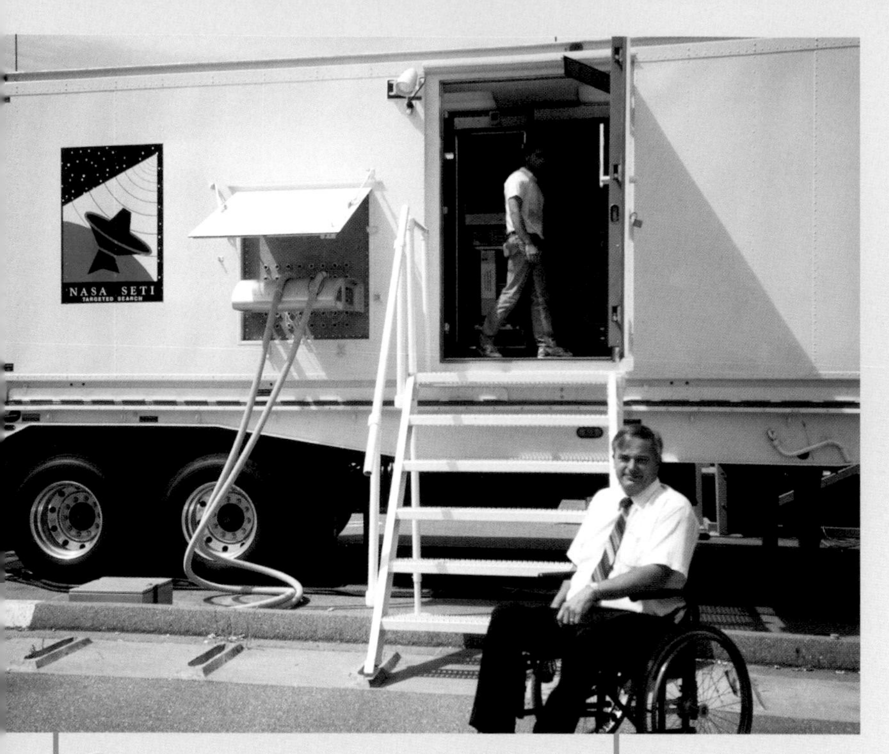

Program Manager David Brocker sits outside the 'MRF'– a container vehicle housing sophisticated receiving equipment – prior to the first SETI search at Arecibo. The poster boasting 'NASA SETI' had to be officially changed to 'NASA HRMS', an acronym for the less controversial 'High Resolution Microwave Survey'.

Two years later – and in front of the same MRF – Project Leader Jill Tarter unveils the privately funded 'Phoenix' search outside the Parkes Telescope in New South Wales, Australia.

Arecibo radio telescope began listening in to signals from the stars.

A year later, the search was dead – killed off by a Nevada senator called Richard Bryan, who derided it as 'a great Martian chase'. Amazingly, he managed to convince a majority of his colleagues in Congress to support him. A *Boston Globe* editorial hit the nail on the head when it opined: 'It proves one thing, and one thing only. That there is no intelligent life in Washington.'

Jill Tarter despairs at the misunderstandings that surround SETI. 'There are many wonderful things about the job I get to do, but there are many frustrating things as well. One is the lack of critical thinking skills among the general population, particularly in the United States. We have worked very hard to set up a scientific research programme that will be credible, verifiable and repeatable. And to have people not appreciate the difference between that and someone who says, "You know, I had an alien in my bedroom last night. 'Course, you can't see him, but I can."'

NASA has no plans to reinstate its SETI programme. Scott Hubbard, the former 'Mars Tsar' at NASA Headquarters, defends the decision. 'While it's maybe financially a small investment, it's also – in many people's view – a real long shot. You might search for 1,000 or 10,000 years and not see anything, given the number of potential objects in the sky. So we remain intellectually interested in life in the Universe, and we wait to see if there's any movement in Congress or the Administration to put this back.'

The SETI team, however, were not prepared to wait around. They rapidly turned themselves into fund-raisers. Within two years of the NASA closure, they had started a new initiative – the privately funded Project Phoenix, controlled from the newly created SETI Institute. And it's going well – although it still has to deliver the phone call from ET. 'We have an observing run coming up again at Arecibo, partnered with Jodrell Bank,' enthuses Jill Tarter. 'And now we've become middle-aged – SETI is forty – it's time to build our own telescope.

'We and the University of California at Berkeley are involved in plans to build a dedicated facility to do radio astronomy and SETI at the same time, all the time. It's going to be something like a thousand small dishes in northern California, covering an area of ten thousand square metres. We call it the 1HT – the One-Hectare Telescope – and in the time it takes Project Phoenix to look at a thousand stars, it'll maybe be getting on for a million.'

SETI researchers are now also looking for artificial laser signals from space. 'We've developed very fast photo-detectors over the past few decades,' says Tarter, 'which could look for fast optical pulses. For a very brief period of time – say, a billionth of a second – the technology could actually outshine its star. Since the transmitter isn't on for very long, the energy budget they have to pay is minimised. So optical SETI becomes an attractive alternative.'

In the nine years we've known the SETI scientists, the big advance that has taken place is the detection of planets around other stars. So does this make Tarter optimistic about getting a signal? 'The extrasolar planetary systems are significantly different from our own. We now know that 5 per cent of stars like the Sun will have "hot Jupiters" in very close orbits around their star. We don't really know about the other 95 per cent, and we don't yet have the technology to find solar systems like our own.'

'We've found about ten Jupiters now that orbit their star very close,' recounts Geoff Marcy, 'closer than Mercury orbits the Sun. And all of the extrasolar planets that orbit further out have highly elongated, elliptical orbits. If there are many planets orbiting the same star, they can bend each other's trajectories – so violently, occasionally, that they throw themselves right towards each other, sometimes crashing like bumper cars. And sometimes they get shunted into really wacky orbits.

'And this yields a new question, which is sort of like looking in the mirror. If other planetary systems have their planets in wacky orbits, why is our Solar System somehow immune to this? The only answer I've ever heard – and one I've been thinking about quite a bit – is that we have something else special here, and that is the planet Earth harbours life.

SETI researchers are now also looking for artificial laser signals from space

'Our Earth is in a near-circular orbit, and we enjoy nearly constant temperatures despite the seasons. If the Earth were in an elliptical orbit, then we'd find ourselves on a very hot planet when the Earth was close to the Sun, and a very cold planet when it was far away. So it may be no coincidence that our Solar System is characterised by circular orbits. Were it not so, we wouldn't be here asking the question.'

The goal of the planet-seekers is to find systems with small worlds like the Earth or Mars in near-circular orbits. And Geoff Marcy is starting to go down that road. 'On the latest Keck observing run, we did something we've never done before. The innocuous, unnamed star HD209458 is the only one we know that has a planet that crosses in front of it. We were hoping to see absorption of light from the star by the planet, and analyse the gases in its atmosphere. We've never been able to find out the stuff of which a planet is made. So we're very, very excited by this. We haven't had a chance to analyse the data yet

Father of SETI, Frank Drake, by an early radio telescope in Silicon Valley. Drake is hugely optimistic that there are many intelligent civilisations in the Universe, and believes that some of them may be immortal.

– we just got it twenty hours ago.'

Jill Tarter anticipates the day when a similar technique can be flown in space, to detect even smaller worlds. 'You look for the effect of a tiny planet like the Earth passing in front of a star and blocking off some of its light. This has to be done from space, because the precision is vastly greater than you can get on the ground – and although such missions haven't been funded, they're on the drawing-board. And then maybe a decade, two decades from now, we can look forward to using an instrument that can make single pixel images of a terrestrial planet.'

Tarter's colleague at the SETI Institute, Chris Chyba, also crystal ball-gazes two decades into the future. 'We're going to have catalogues of hundreds and possibly thousands of extrasolar planets, and there's nobody better than Geoff Marcy to tell you about that. And new technologies are going to show up, including space, which will accelerate the discovery-rate even beyond what's happening now.

'We're going to get to know whether our Solar System is typical or rare or something in between. I think we're going to discover that it is neither rare nor typical, that it is one of a whole suite of possible solar systems. But the point we're going to know is whether the Earth is common or rare, or unique. I expect we'll discover that the Earth is not unusual, but we don't know. We'll know in the next ten or twenty years.'

This, of course, begs the question as to whether the other Earths out there could harbour life – which, for practical reasons of communication, means that it must be intelligent and technologically literate. How likely is it that microbial life (nicknamed 'green slime' by biologists) might develop into intelligent life? 'Well now, if I knew the answer to that I'd be very famous,' laughs Frank Drake. 'But I think that the biological potentiality is there to produce a host of intelligent creatures. If the Earth is at all typical, it says to me that every planet whose star has a lifetime of 5,000 to 10,000 million

years has a good chance of producing an intelligent species. Another 1,000 million years, you know... that's a lot of time. We humans just aren't able to think in those terms very well.'

'I don't think we have a good answer to that question,' ponders Chris Chyba. 'The search for extraterrestrial intelligence remains fundamentally an empirical problem. We're not going to find answers on the basis of one example on Earth. We have to do the search – the search for intelligence, the search for micro-organisms within our own Solar System. And we should look at the atmospheres of extrasolar planets once we have the technology to do so, and try to find planets that have chemical disequilibria in their atmospheres that is consistent with life.'

British independent scientist and Gaia-guru Jim Lovelock agrees, but can't stifle a rueful grin. 'The piquant thing I find about NASA's activities at the moment is that in the 1960s and early 1970s they rejected my atmospheric analysis method for life detection. They wouldn't look at it for Mars, but that's exactly the method they're proposing to look for life on extrasolar planets. And they give no back credit – they're a bit naughty that way.'

'The Universe is so old that if aliens evolved anywhere – and they should have done, because life evolved so quickly on the Earth – then we should see them, we should see their ruins'

Not everyone is sanguine about the chances of finding intelligent life elsewhere in the Universe. Sci-fi writer Stephen Baxter is pessimistic – to his own dismay. 'I'm working through a series of books now about the Fermi Paradox, which says that the Universe is so old that if aliens evolved anywhere – and they should have done, because life evolved so quickly on the Earth – then we should see them, we should see their ruins. I mean, the Universe should be like Britain, with layers of history. Roman ruins overlaid by Saxon ruins, overlaid by Norman ruins and so forth. And we see just nothing – not a sausage or a squeak as far back as we look in time.

'If I had to put my money on the right answer, it's surely Occam's Razor – you know, the simplest answer. We don't see them, so they probably don't exist. However, that's an awful thought –

The four matching domes of the VLT – the world's most advanced telescope – are perched on a remote mountain ridge at Paranal in Chile. The recently completed telescopes have already started to detect planets around other stars.

7 April 2001: the latest probe to the Red Planet, Mars Odyssey, lifts off from the Kennedy Space Center atop a Delta launcher. Optimism about finding some kind of life on Mars has spilled over into the hope of detecting it on planets around other stars.

that we on this tiny rock are alone, and there's nobody to ask what the hell's going on. That's just a desolating thought. I hope it's wrong – but that's where my money is.'

Geologist-cum-astrobiologist Bruce Jakosky doesn't agree. 'We've been doing SETI for forty years or so, and we haven't found life. Maybe radio is such a flash in the pan, galactically speaking, that nobody uses it any more. We might need to use radio for fifty years, and then move on to something else. I don't think we can conclude that there is no life. I don't think we can conclude anything except that we don't have enough data.'

We talk to Jakosky in a spacious room on the campus of the University of Colorado at Boulder. Like many astrobiologists, he loves the wide perspectives and the philosophical concepts that his work brings him. Half of our conversation concerns his passion for reaching out to the public and explaining what guides and motivates science and scientists. He can pinpoint when his direction changed, from straight science to a fascination with its wider implications.

'I was studying Mars,' he recalls, 'and I didn't see how it all connected up. I didn't see what the broader significance of what the geology was like, or what drove the climate. I'd actually started getting bored with the planet until I got interested in the question about life.

'Then you find yourself asking questions like: what does Mars's existence tell us about the Earth's existence? Does Mars have life, and what does that tell us about life on Earth – and vice versa? You very quickly back up to some of the broader-picture questions. And the question of life pretty soon leads you to find out that you're asking the big questions – capital B, capital Q.'

Jakosky's big questions kick off with the origin of the Universe. And can we predict its final demise – will it go on expanding, or even accelerating into the future? He cites a fascination with how galaxies form, where stars and planets come from, and whether there are Earth-like planets out there. 'But at the pinnacle of all this,' he observes, 'it's whether there is life – intelligent life – out there.

'I find that everybody is interested in this question. Everyone has a personal stake in whether there's life elsewhere. It could be microbes on Mars, intelligent life on a planet in another solar system, or UFOs and alien abductions. It means something to everybody.

'One of my students put it very aptly in saying that we're looking for ourselves. I think she meant that we're looking for beings out there with whom we can communicate and so find out more about ourselves. The search for life is no different from, in my mind, things we do to understand who we are. Exploring the arts, exploring literature. People talk about using novels to explore the human condition. We're exploring what it means to be human – exploring the mind, philosophy, all of these are different ways of understanding who we are.

'We ask the big questions because – I think – we relate to the Universe, and the Universe relates to us'

'We ask the big questions because – I think – we relate to the Universe, and the Universe relates to us.'

Jakosky's observations of how we relate to the Universe put Mars in its human context. It is the nearest world to Earth that may harbour life, however primitive; and we know that we can get there. It is a metaphor for our future exploration of the Universe. Human beings will

Mars, swept by dust-storms, is here captured in a sensational new image from the Hubble Space Telescope. It will be our home in the future – and later, our springboard to the stars.

The Mars Society's Bob Zubrin agrees. 'People will go to Mars. They will endure all kinds of hardship for that kind of freedom. But at a certain point, Mars will be tamed – a place where the rules will have been written. Then those kinds of bold spirits will have need to go out further. I think that humanity will continue to expand for a fundamental reason. And that's because humans – or certain humans – will always have a deep drive to be the creators of their own world, and not just its inhabitants.'

'Let's think about what we humans have done,' suggests Geoff Marcy; 'what we've been compelled to do as a species, for reasons that are psychologically unconscious to us. The Polynesians left in tiny canoe-like boats for parts unknown, and discovered Hawaii. They couldn't have had a purpose when they stumbled across it – they were just compelled to go out and see what was there.'

undoubtedly travel to Mars in this century. But beyond that?

'To go beyond Mars, the first question you would pose is: why? What scientific or national or even human imperative are you following?' asks NASA's Scott Hubbard. 'But if there's some extraordinary breakthrough that makes access to interplanetary or even interstellar space affordable in some sense, then there's going to be a group of explorers who will want to go – I have no doubt of that.'

'In the sixteenth century the Europeans went across the Atlantic for those ridiculously long voyages. People died by the dozen, but they did it anyway.

'And I think that we humans – some of us, anyway – will feel a compulsion to travel across the ocean of space towards whatever destination brings a twinkle to our eye. It'll be Mars in the near future, and it'll be the planets around other stars in the distant future.'

INDEX

Illustrations and publications are shown in italics

D

E

F

G

H

PICTURE CREDITS

pp.1, 2–3, 4–5 and 6–7 © NASA; p.3 © NASA/Science Photo Library.

CHAPTER 1

pp.8–9, 10 and 25 © NASA/Science Photo Library; p.12 © The Natural History Museum, London; p.14 © NASA; p.16a © KAL; p.16b, 16c and 16d © John Frost Newspapers; p.18a © NASA/JSC; p.18b courtesy of Nadine G. Barlow and NASA; p.20 © Antarctic Search for Meteorites (ANSMET) program/W.A. Cassidy; p.21 © Scientific American; p.22 © Kathie Thomas-Keprta.

CHAPTER 2

pp.28–30 and 37 © The Kobal Collection; p.31 and 36 © Mary Evans Picture Library; p.32 © AKG London/ Erich Lessing; p.34 © Science Photo Library; p.35 Peter Menzel/Science Photo Library; p.39 © John Frost Newspapers.

CHAPTER 3

pp.40–1, 42, 44, 45, 46, 49, 50, 52, 56 and 58–9 © NASA; p.43 and 55 © NASA/JPL/Malin Space Science Systems; p.44–5 © NASA/Science Photo Library; p.47 © William K. Hartmann.

CHAPTER 4

pp.60–1 and 79 © Alfred Pasieka/ Science Photo Library; p.62 © Jonathan Trent and Susanne Johansen Trent; p.63 © NASA/ JPL/Malin Space Science Systems; p.64 © James King-Holmes/ Science Photo Library; p.66 © J.C. Revy/ Science Photo Library; p.67 © Victor Habbick Visions/ Science Photo Library; p.68 © Hencoup Enterprises; pp.68–9 © Tony & Daphne Hallas/ Science Photo Library; p.69 © European Southern Observatory; p.71 © Kathie Atkinson/ Oxford Scientific Films; p.73 © B. Murton/Southampton Oceanography Centre/Science Photo Library; p.74 © NASA; p.76 Jonathan Trent.

CHAPTER 5

pp.80–1 © NASA/Science and Society Picture Library; p.82, 83, 84, 85, 90, 91 and 96 © NASA; p. 87 © Space Frontiers; p.88 © Hencoup Enterprises; pp.94–5 © US Geological Survey/Science Photo Library.

CHAPTER 6

pp. 100–1 © Novosti Press Agency/ Science Photo Library; p.102 © Mary Evans Picture Library; pp.104, 105 and 107 © US Naval Observatory; p.109 © Novosti (London); p.110 © The Advertising Archive; p.112 © David A. Hardy; p.113 © NASA/ Galaxy; p.114 © NASA; p.115 © NASA/Science Photo Library.

CHAPTER 7

pp.116–7 © NASA; p.118 © NASA/ Science Photo Library; p.119 © NASA/ JPL; pp.120–1 and 129 (b) © Novosti (London); p.123 (a) © MSSS/JPL/NASA/ Science Photo Library; p.123 (b), 126 and 127 (a) © NASA/JPL/Malin Space Science Systems; p.127 (b) and 128 Detlev Van Ravenswaay/Science Photo Library; p.129 (a) © Hencoup Enterprises; p.133 © Tony Morrison/ South American Pictures.

CHAPTER 8

pp.134–5, 143, 146–7, 150 and 154 © NASA; pp.136–7 and 138 © NASA/ Science Photo Library; p.139, 141 and 142 © NASA/JPL/Malin Space Science Systems; p.144 © Peter Menzel/Science Photo Library; p.152 © University of Leicester.

CHAPTER 9

pp.156–7 © NASA; p.159 © M.I. Walker/ Science Photo Library; p.161 © The Bridgeman Art Library; p.162 © Michael Carroll; p.163 and 164 (a) and (b) © NASA, Haughton-Mars Project 2001; p.166 © Robert Zubrin.

CHAPTER 10

pp.168–9 © US Geological Survey/ Science Photo Library; p.170, 171 and 182–3 © NASA; p.172 © The Advertising Archive; pp.174–5 © John Frassanito and Associates/ NASA; p.176 © NASA/ Science Photo Library; p.178 © Novsoti (London); p.180 © Science Photo Library.

CHAPTER 11

pp.186–7 and 198 © NASA; p.188 © Mark Garlick/Science Photo Library; p.189 © NASA/Science Photo Library; p.191 © The Advertising Archive; p.192 and 196 © Andrew Stewart; p.193 and 194 © Mark Garlick.

CHAPTER 12

pp.200–1 © Pat Rawlings/NASA; p.202 © Magrath Photography/Science Photo Library; p.203 and 218 © NASA and The Hubble Heritage Team (STScI/Aura); p.205 © European Southern Observatory; p.206 © NASA; p.208 and 216 © NASA/ JPL; p.209 © NASA/JPL/University of Arizona; pp.211, 212, 214 and 214–15 © Hencoup Enterprises.